Atom Optics with Laser Light

Laser Science and Technology
An International Handbook

Editors in Chief

V.S. LETOKHOV, Institute of Spectroscopy, Russian Academy of Sciences, 142092 Moscow Region, Troitsk, Russia

C.V. SHANK, Director, Lawrence Berkeley Laboratory, University of California, Berkeley, California 94720, USA

Y.R. SHEN, Department of Physics, University of California, Berkeley, California 94720, USA

H. WALTHER, Max-Planck-Institut für Quantenoptik und Sektion Physik, Universität München, D-8046 Garching, Germany

Atom Optics with Laser Light

V.I. Balykin and V.S. Letokhov

Institute of Spectroscopy
Russian Academy of Sciences
Troitzk, Moscow Region
142092 Russia

Routledge
Taylor & Francis Group

LONDON AND NEW YORK

First published 1995 by Harwood Academic Pubishers GmbH.

Published 2018 by Routledge
2 Park Square, Milton Park, Abingdon, Oxon OX14 4RN
52 Vanderbilt Avenue, New York, NY 10017

Routledge is an imprint of the Taylor & Francis Group, an informa business

British Library Cataloguing in Publication Data

Balykin, V. I.
  Atom Optics with Laser Light. – (Laser
  Science & Technology Series, ISSN
  0899–2711; Vol. 18)
  I. Title II. Letokhov, V. S. III. Series
  621.366

ISBN 13: 978-3-7186-5697-4 (pbk)
ISBN 13: 978-1-138-45581-8 (hbk)

CONTENTS

Introduction to the Series

Almost 30 years have passed since the laser was invented; nevertheless, the fields of lasers and laser applications are far from being exhausted. On the contrary, during the last few years they have been developing faster than ever. In particular, various laser systems have reached a state of maturity such that more and more applications are seen suffusing fields of science and technology, ranging from fundamental physics to materials processing and medicine. The rapid development and large variety of these applications call for quick and concise information on the latest achievements; this is especially important for the rapidly growing inter-disciplinary areas.

The aim of *Laser Science and Technology – An International Handbook* is to provide information quickly on current as well as promising developments in lasers. It consists of a series of self-contained tracts and handbooks pertinent to laser science and technology. Each tract starts with a basic introduction and goes as far as the most advanced results. Each should be useful to researchers looking for concise information about a particular endeavor, to engineers who would like to understand the basic facts of the laser applications in their respective occupations, and finally to graduate students seeking an introduction into the field they are preparing to engage in.

When a sufficient number of tracts devoted to a specific field have been published, authors will update and cross-reference their pages for publication as a volume of the handbook.

All the authors and section editors are outstanding scientists who have done pioneering work in their particular field.

V.S. Letokhov
C.V. Shank
Y.R. Shen
H. Walther

Main Notations

d	electric dipole moment
D_{ij}	diffusion tensor
D_{ii}^a	anisotropic diffusion tensor
D_{ii}^d	directed diffusion tensor
$E(r)$	electrical field strength
F	force
F_{RAD}	radiation force
F_{GR}	gradient force
F_{LP}, F_{sp}	light pressure force
F_{RT}	retarded (stimulated) force
g	gravitational constant
g_{at}	atomic degeneracy
G	resonant saturation parameter
G_0	resonant saturation parameter in a standing wave
h	Planck's constant
$\hbar = h/2\pi$	
$H(r)$	magnetic field strength
I	intensity of light wave
I_s	saturation intensity
k	wave number
k_B	the Bolzmann constant
k_{Br}	de Broglie wave number
l	length of resonator
M	mass of particle
n_2	population of excited state
p	momentum of particles
P	laser power
R	recoil energy
s	off-resonance saturation parameter
T	temperature

$T(x,y)$	phase transmission function
U	potential
v	velocity
v_{long}	longitudinal atomic velocity
v_{rec}	recoil velocity
W_i	internal energy of an atom in a quantum state
2γ	natural width of absorption line
μ	magnetic moment
λ	wavelength of light
λ_{Br}	de Broglie wavelength
η	damping coefficient
Q	rate of photon scattering by atom
Θ	transmission coefficient through potential barrier
θ	phase of electrical field
ρ	distance between atom and laser beam axis
ρ_0	radius of a laser beam
σ	radius of atomic beam
σ_0	waist of atomic beam
σ_{ch}	radius of atomic beam in image plate due to chromatic aberration
σ_s	radius of atomic beam in image plate due to spherical aberration
σ_{diff}	radius of atomic beam in image plate due to diffraction aberration
τ	time of collimation
τ_{RD}	time of collimation by retarded force
$\psi(r,t)$	wave function
ω	frequency of light
ω_R	the Rabi frequency
Ω	detuning of field frequency from absorption frequency

1. INTRODUCTION

The term of *atom optics* is due to the natural analogy with *light optics* or the optics of photons. Light optics is based on the two fundamental principles: (a) the wave properties of light and (b) the electromagnetic interaction between light field and matter or, in other words, between light and bound charged particles (electrons or ions) in a medium. Owing to this interaction, the light field can be reflected by the medium or diffracted by it, or else light can propagate through the medium with some velocity other than the velocity of light in a vacuum, and so on (Born and Wolf, 1984).

1.1 TYPES OF MASSIVE-PARTICLE OPTICS

According to the de Broglie theory, wavelike properties are associated with any particles of matter, and the de Broglie wavelength is defined by the fundamental relation

$$\lambda_{Br} = h/p = h/Mv \qquad (1.1)$$

where h is Planck's constant and p, M, and v are the momentum, mass, and velocity of particle, respectively. The wave properties of massive particles were verified in experiments on the diffraction of electrons and used in the first light optics analog for particles — *electron optics* (Grivet, 1972). Electron optics is based on (a) the wave properties of electrons and (b) the electromagnetic interaction between moving electronic charge and electrical and magnetic fields of appropriate configuration (Grivet, 1972). The most familiar application of electron optics is electron microscopy (Ruska, 1980).

Another light optics analog is *neutron optics* based again on (a) the wave properties of ultra cold neutron and (b) the interaction between neutrons and atomic nuclei, which can be described by means of what is known as the optical potential [Sears, 1989]. As distinct from electron optics we deal here with more massive particles (ultra-cold neutrons) whose wave properties manifest themselves at low temperatures; [Shapiro, 1976]. The effect of gravitation and low intensity of ultracold neutron sources make experiment in neutron optics more complex than in electron optics. Nevertheless neutron interferometers [Bonze and Rauch, 1979] and microscopes [Shutz et al., 1980; Frank, 1987, 1991] have already been successfully realized.

The next natural object are neutral atoms or molecules. The wave properties of atoms and molecules and various types of their interaction with matter and electromagnetic fields (from static to optical) make it possible to implement *atom* and *molecular optics*. It is precisely the great variety of methods for exerting effect on an

1

atom (or molecule) possessing a static electrical or magnetic moment, a quadruple moment and optic resonance transitions (or a high frequency dipole moment) that form the basis for several possible ways to realize atomic (molecular) optics. Let us consider them briefly.

1.2 METHODS OF REALIZATION OF ATOM OPTICS

The known methods to implement atom optics (atomic–optical effects) can be classed in the following three categories:

(a) methods based on the interaction between atoms and matter;

(b) methods based on the interaction between atoms having a magnetic or electrical dipole moment and a static electrical magnetic field of a suitable configuration;

(c) methods based on the resonance (or quasiresonance) interaction between atom and a laser field.

The first experiment on atom optics realized by method (a) and (b) were succesfully conducted almost a century ago. The advent of tunable laser allowed the possibility to demonstrate atom optics based on the atom-light interaction. It is exactly this type of atom optics that the present review is devoted to. However, for the sake of generality of the physical picture, it seems advisable to recall briefly the milestones in all approaches to atom (molecular) optics.

1.2.1 Interaction Between Atoms and Matter

In his classical monograph, Ramsey (1956) (Chap. 2, Sect. 5) considered the mirror reflection and diffraction of molecular beams on the surface of a solid. According to Ramsey (1956), for mirror reflection to occur, it is necessary that the following two conditions be satisfied.

(a) The projection of the height of surface irregularities on the direction of molecular beam must be shorter than the de Broglie wavelength. Recall an example from light optics: smoked glass is a poor reflector in the case of perpendicular incidence and a good reflector in the case grazing incidence. If δ is the average height of surface irregularities and ϕ is the grazing angle of incident beam, the above requirement may be expressed as (Fig. 1a).

$$\delta \sin \phi < \lambda_{Br} \tag{1.2}$$

(b) The average residence time of the particle on the surface must be short. In this case, the state of reflected particles will be the same as that of incident particles.

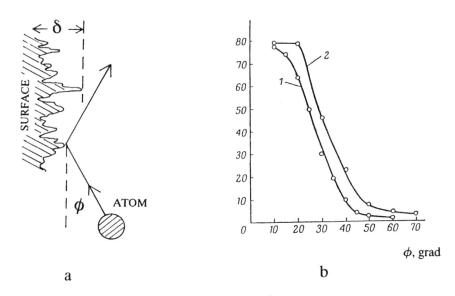

a b

Fig. 1 Reflection of atom in grazing incidence upon the surface of a solid: (a) requirement for the surface roughness δ, the glancing angle of the incident beam ϕ and the de Broglie wavelength; (b) reflectivity of a beam of He atoms from the surface of LiF crystal at two different temperatures (1 — 295 K; 2 — 100 K), (Estermann and Stern, 1930).

The roughness of most thoroughly mechanically polished surfaces is of the order 10^{-5} cm, whereas the de Broglie wavelength of hydrogen at 300 K amounts to 10^{-8} cm. Therefore, according to (1.1) and (1.2), the condition for the reflection has the form $\phi < 10^{-3}$ rad.

It was more than 50 years ago that they managed to observe a 5% reflection of hydrogen beam from polished bronze mirror at the grazing angle of $\phi = 10^{-3}$ grad (Krauer and Stern, 1929). Cleaved crystal surfaces are much more smoother. The thermal vibrations of the crystal lattice limits the roughness of the surface to about 10^{-8} cm. In that case a beam of He atoms should undergo reflection at grazing angles less than 20–30 grad. This was confirmed in the experiments (Estermann and Stern, 1930) with He atoms and LiF crystal. (Fig. 1b). The temperature dependence of the grazing angle marking the onset of simple reflection of atoms bears witness of the fact that thermal vibrations have an effect on the surface roughness of crystal.

Experiment on the simple reflection of atoms at the surface of condensed medium continue to draw investigator's attention. Recall the experiments on the reflection of ^{4}He atoms grazing the surface of liquid ^{4}He [Nayak et al., 1983] and thermal Cs atoms grazing a polished glass surface [Anderson et al., 1986].

The first experiment aimed at observing the diffraction of atoms by a cleaved crystal surface acting as a two-dimensional, plane grating were conducted by Stern [1929] and the results of detailed research into this phenomena were presented in [Frish and Stern, 1933]. The diffraction of atoms by a fabricated periodic structure (a slotted membrane) with a much more grating period was observed in the work reported in [Keith et al., 1988].

The effect of quantum reflection of ^{4}He and ^{3}He beams at a liquid-helium-vacuum interface was successfully used to focus hydrogen atoms with a concave mirror [Berkhout et al., 1989], and the authors of [Carnal et al., 1991] were successful in conducting an experiment on focusing a beam He atoms by means of a zone plate. In [Ekstrum et al., 1992] was proposed the several optical elements based on the microfabricated structures.

Atomic interferometry based on the microfabricated structures was realized in two elegant experiments: The atomic Young's two slit interferometer [Carnal and Mlynek, 1991] and the atomic Michelson interferometer [Keith et al., 1991].

1.2.2 Interaction Between Atoms and Static Electric and Magnetic Fields

Some elements of the optics of atoms and molecules, based on the interaction between spatially nonuniform static magnetic or electrical fields and the magnetic or electrical dipole moment of the particles, have long been known and used fairly successfully in experimental physics. An excellent review on the early experiments in this field was presented by Ramsey [1956].

In the presence of a magnetic or electric field, the quantum state of atom or molecule are shifted, the shift depending on the initial quantum state of the particle and the field strength (the Zeeman and Stark effects). In the adiabatic approximation (the field varies in the time and space not very rapidly, the particles move slower enough), the internal state of particles follows the field strength variation, or, in other words the particles remains at one and the same quantum sublevel whose energy W depends on the field strength.

In the adiabatic approximation, the motion of the center of mass of a neutral particle with a mass M obeys the Schrödinger equation for the wave function $\psi(r, t)$:

$$ih[\partial \psi_i(\mathbf{r},\ t)/\partial t] = \left\{ -\hbar^2/2M)\nabla^2 + W_i(\vec{r}\,) \right\} \psi_i(\mathbf{r},\ t) \qquad (1.3)$$

where $W_i(\mathbf{r})$ is the internal energy of the particle in the quantum state i at the point $\mathbf{r}$ which depends on the electrical field strength $E(\mathbf{r})$ or the magnetic field strength $H(\mathbf{r})$.

Magnetic Interaction. In the simple case of a constant magnetic moment μ, the effective potential energy W of an atom or a molecule in an external magnetic field of strength H is given by

$$W = -\mu H = -\mu_{eff} H \qquad (1.4)$$

where μ_{eff} is the projection of μ on the direction of **H**. It follows from the relationship between force and potential energy that the force acting on the atom or molecule is

$$\mathbf{F} = -\nabla W = -(\partial W/\partial H)\nabla H = \mu_{eff}\nabla H \qquad (1.5)$$

A particle in nonuniform magnetic field is acted upon by the force directed along the field strength gradient.

The authors of [Friedburg and Paul, 1950, 1951; Korsynskii and Fogel, 1951; Vanthier, 1949] proposed to use nonuniform magnetic field to focus molecular beams issuing at different angle from the source.

Figure 2 shows the configuration of the focusing magnetic field used by Frieburg and Paul [1950, 1951]. The method was extended by Benewitz and Paul [1954] to atoms whose magnetic moments depend on the strength of the external magnetic field.

The focusing properties of a magnetic lens depends on the magnetic sublevel of atoms. That was successfully used by Ramsey and co-workers to create the hydrogen maser [Goldenberg et al., 1960; 1962]. The hydrogen atoms in the state $F = 1, M = 0$ were focused in a small hole in the wall of the storage cell and accumulated there, while the atoms in the lower atoms state $F = 0$ are defocused.

Electrical Interaction. Since the energy of atom or molecule in an electrical field depends on the strength of the latter, it then can be presumed, by analogy with (1.4, 1.5), than the atom or molecule possesses an effective dipole moment given by

$$\mu_{eff} = -(\partial W/\partial E) \qquad (1.6)$$

The force acting on an atom or a molecule in a nonuniform electrical field is defined by the following expression similar to (1.5):

$$\mathbf{F} = \mu_{eff}\nabla E = \mu_{eff}(\partial E/\partial z) \qquad (1.7)$$

where the direction of the field strength gradient is taken to be the z-axis.

Paul and co-workers [Benewitz et al., 1955] created focusing electrical fields for a beam of polar molecules. The electrical focusing of a beam of molecules in a certain (excited) quantum state was used by Townes in developing the NH_3 maser [Gordon et al., 1955]. The hexapolar electrical field configuration (as in Fig. 2) posesses not only focusing properties, but also selectivity with respect to the quantum state of the molecule, for the quantity μ_{eff} depends on its quantum numbers J, K, and M.

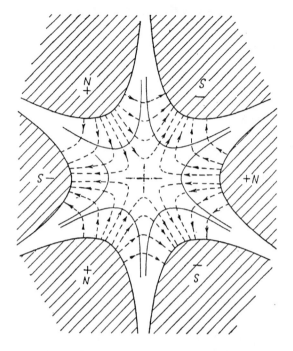

Fig. 2 Focusing magnetic field of Friedberg and Paul [1950]. Illustrated are the lines of forces (dashed) and equipotential lines (solid) of magnetic lens.

This latter property was successfully used in experiment on molecular dynamics with a beam of molecules in a specified quantum state, including the experiments on the orientation of molecules [Bernstein, 1982].

When speaking of the optics of atomic or molecular beams, we almost always mean their focusing, for it is exactly this effect that has found practical application. But one can also speak electrical or magnetic mirrors and gratings for slow moving neutral atoms and molecules [Opat et al., 1992].

1.2.3 Interaction between Atoms and Light Field

Atoms or molecules having no static magnetic or electrical dipole moment cannot change their mechanical trajectory in a static magnetic or electrical field. However, new possibilities are being opened up for particles, based on the induction in them of a high frequency (optical) electrical dipole moment in a quasiresonant or resonant laser light field. Before the advent of the laser, it was only possible to induce microwave

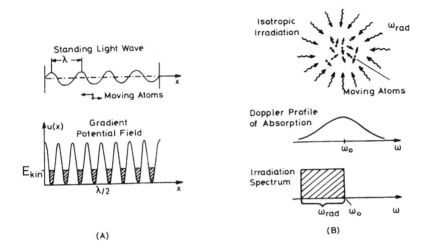

Fig. 3 Illustrating the idea of (A) channeling or localization of atoms (Letokhov, 1968), and (B) cooling of atoms and ions (Hänsch and Schawlow, 1975; Wineland and Dehmelt, 1975) by laser radiation.

transitions in atoms and molecules, which allowed one to alter efficiently their quantum state, and thereby the character of motion in external spatially nonuniform electric or magnetic field [Ramsey, 1956].

An atom in quasiresonant laser field acquires a high-frequency polarizability, and if the intensity of laser field is spatially nonuniform, the atom is acted upon by the *gradient* (*dipole*) force (Askarian, 1962; 1972). For example the gradient force in a standing laser light wave may cause the channeling of atoms moving along the wave front, Fig. 3(a) [Letokhov, 1968]. The gradient force was successfully used by Ashkin and co-workers to focus an atomic beam [Bjorkholm et al., 1978].

In the optical region of the spectrum, the recoil effect resulting from atom light interaction is significant. This effect was predicted by Einstein [Einstein, 1909; 1917] as far as back as 1909, was experimentally corroborated by the slight deflection of a beam of sodium atoms scattering the resonant radiation of D line of Na [Frisch, 1933].

An intense laser radiation tuned to the resonance with some allowed dipole transition in an atom can make it re-emit millions of photons, and as a consequence a radiation can exert a substantial effect on the atomic velocity and mechanical trajectory. Hänsch and Schawlow (1975) proposed to use the resonance force due to spontaneous re-emission of photons for cooling neutral atoms (Fig. 3b) and Wineland and Dehmelt (1975) for cooling ions in an electromagnetic trap.

The gradient and spontaneous forces are at the root of a great many of experiments on controlling the motion of atoms by means of light, which was already considered in the reviews [Ashkin, 1980; Letokhov and Minogin, 1981; Dehmelt, 1983; Stenholm, 1986; Wineland and Itano, 1987; Phillips et al., 1988; Cohen-Tannoudji and Phillips, 1990; Chu, 1991; Cohen-Tannoudji, 1991]; special issue of scientific journals [Meystre and Stenholm, eds., 1985; Chu and Wieman, eds., 1991;], monographs [Minogin and Letokhov, 1978; Kazantsev et al., 1991] and conference reports [Moi et al., eds., 1991]. The problem of atom optic have been recently discussed in the short reviews [Balykin and Letokhov, 1989], [Pritchard, 1991; Pritchard and Oldaker, 1990] and a special issue [Mlynek, Balykin and Meystre, eds., 1992].

The aim of the present review is to examine in detail only one trend in atom optic, namely, *laser-induced atom optics*. But before proceeding to discuss these special problems, it may be worth while to describe briefly the nature of the radiation forces acting on an atom in a laser field. A more detailed and rigorous discussion of these forces can be found in the lectures of Cohen-Tannoudji (1991).

2. RADIATION FORCES FOR MANIPULATION OF ATOMS

Under the name *radiation force* will be understood the total force arising upon interaction between laser light and an atom. Depending on the spatial and temporal structure of light field, its strength and wavelength, the radiation force may be a very complex function of the atoms position and velocity. But, since all the known studies on the application of the radiation forces have been carried out mainly using three types of light fields, namely, a plane traveling light field, a Gaussian laser beam and a standing light wave, or their combination, we will restrict ourselves to qualitative consideration of these types of fields only.

In this review we will consider the motion of atom under different types of radiation forces. Since the conception of force is a classical one, it is necessary to mention under what condition the motion of atom can be considered as a classical one [Minogin and Letokhov, 1987; Cohen-Tannoudji, 1991].

There are two such conditions. One of them directly follows from the fact that for classical atomic motion the quantum fluctuation in the atomic momentum must be negligibly small compared to the change in the average atomic momentum due to the force. The minimum atomic momentum variation in a laser field is equal to the photon momentum. The change in the average atomic momentum under action of the radiation force is to be considered significant if it breaks of the resonant atomic interaction with the light field. The frequency range within which an atom absorbs the resonant radiation is determined by natural width of absorption line 2γ. The resonance between atom and light is interrupted when the average atomic momentum varies by the value $\delta p \approx M\gamma/k$. If one requires that the value of this atomic momentum δp is larger than the photon momentum, the first condition for a classical atomic motion will be

$$\hbar k \ll M\gamma/k \qquad (2.1a)$$

or

$$R \ll \hbar\gamma \qquad (2.1b)$$

where $R = \hbar^2 k^2/2M$ is the recoil energy.

The second condition for classical motion follows from the fact that the process of absorption — spontaneous emission takes place in time interval of the order of $\tau_{sp} \approx \gamma^{-1}$. Because the classical description of atomic motion cannot allow for such small scale momentum variations, it is necessary to restrict classical time scale to the condition:

$$\delta t \gg \gamma^{-1} \tag{2.2}$$

We could look at the problem of classical description of motion of atom from the point of view of an evolution of the atomic wave packet. The quantum description of the evolution of the atomic wave packet will be close to the classical one if the atom has a well defined position and momentum. The momentum width Δp of wave packet related to the position width Δr by the Heisenberg inequality

$$\Delta p * \Delta r > \hbar \tag{2.3}$$

The force exerted by laser field on the atom varies over distances on the order of laser wavelength λ_1, or larger. The force could be considered classical if the position spread of wave packet is a smaller than the laser wavelength:

$$\Delta r \ll \lambda_1 \tag{2.4}$$

The force also depends on the velocity of atom because of the Doppler shift. The appreciable change of the atomic response to the laser excitation through the Doppler shift will be at the velocity change:

$$\delta v \cong \gamma/k \tag{2.5}$$

It is clear that velocity (momentum) spread Δv of wave packet (for classical description of atomic motion) must be smaller than δv:

$$\Delta v \ll \delta v = \gamma/k \tag{2.6}$$

or

$$\Delta p \ll \gamma M/k \tag{2.7}$$

Equations (2.4) and (2.7) impose upper bound on the momentum and space spread of wave packet which can be in conflict with the Heidelberg inequality (2.3). From (2.3), (2.4) and (2.7) follows one condition for classical description of atomic motion:

$$\hbar k \ll M\gamma/k \tag{2.8}$$

which also coincides with (2.1).

To date theory of atomic motion in laser fields has been developed quite well [see Meystre and Stenholm, eds., 1985; Cohen-Tannoudji, 1991; Chu and Wieman, eds., 1991; Minogin and Letokhov, 1987; Kazantsev et al., 1991].

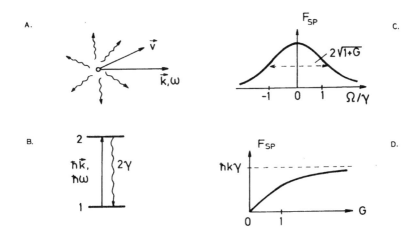

Fig. 4 Origin and properties of light pressure (spontaneous) force F_{sp}: (A) absorption of directed $(\mathbf{k}, \omega)$, photon and isotropic remission of spontaneous photon by atom with velocity v; (B) an ideal two level diagram of resonant interaction of light wave with frequency ω and the quantum transition frequency ω_0 with the radiative damping rate 2γ of the excited state; (C) resonant dependence of F_{sp} as a function of frequency detuning $\Omega = \omega - \omega_0 = \mathbf{k}v$; (D) saturable dependence of F_{sp} as a function of saturation parameter G.

2.1 A PLANE TRAVELING WAVE. LIGHT PRESSURE (SPONTANEOUS) FORCE

Consider a plane wave directed along the z-axis and having its frequency tuned to resonance with the absorption frequency of an atom placed in it (see Fig. 4):

$$\mathbf{E} = \mathbf{e}E_0 \; \cos(kz - \omega t) \qquad (2.9)$$

The atom absorbs laser photons directed along the z-axis and re-emits spontaneous photons symmetrically in all directions. As result of these processes, a radiation force is determined by the product of the momentum of a photon and the rate of photon scattering.

$$F_{LP} = \hbar k Q \qquad (2.10)$$

where $Q = 2\gamma n_2$ is the rate of photon resonant scattering by atom, and n_2 is a steady state population of exited state.

The atom is acted upon in the direction of the wave by radiation force whose maximum magnitude is given by the product of photon momentum $\hbar k$ by the photon scattering rate Q, i.e., $F_{max} = \hbar k \gamma$, where $k = 2\pi/\lambda$. If the atom is not in exact resonance with laser radiation and has a velocity v projected onto the z-axis in the direction of light wave, a steady state population of exited state depends on the projected velocity and on the detuning of the field frequency ω from the absorption frequency ω_0 of the atom:

$$n_2 = \left[G/2(1 + G + (\Omega - kv)^2/\gamma^2) \right] \qquad (2.11)$$

where $G = I/I_s$, I is the intensity of the wave, I_s is the saturation intensity of the atomic transition, and $\Omega = \omega - \omega_0$ is a detuning of field frequency from an absorpsion frequency.

From (2.11) and (2.12) follows the expression for the radiation force in a traveling wave and which is called a *light pressure force* (Ashkin, 1970):

$$F_{LP} = \hbar k \gamma \left[G/(1 + G + (\Omega - kv)^2/\gamma^2) \right] \qquad (2.12)$$

The force has a Lorentzian dependence on the velocity of atom (Fig. 4c) and reaches its maximum value at exact resonance. The ultimate value of the force is limited by the saturation effect to F_{max}. The acceleration of an atom under the action of this force reaches a magnitude of 10^8 cm/sec^2, which is 10^5 times greater than the acceleration of the earth gravity.

Under optimum condition, dissipative light pressure force (2.12) allows the atoms to be cooled to the temperature of the so called Doppler limit [Letokhov et al., 1977; Wineland and Itano, 1979]:

$$T_{min} = \hbar\gamma/2k \qquad (2.13)$$

2.2 TRAVELING GAUSSIAN LIGHT BEAM. GRADIENT (DIPOLE) FORCE

In the case of a Gaussian light beam of spatially nonuniform distribution of amplitude and phase the electrical field is:

$$E = E_0(r) \cos(kr - \omega t), \quad E_0(r) = E_0 \exp(-r^2/2\rho^2) \qquad (2.14)$$

In such field an atom is acted upon the radiation force F_{RAD} which can be expressed as the sum of the light pressure force F_{LP} and what is known as the gradient force F_{GR} (Fig. 5):

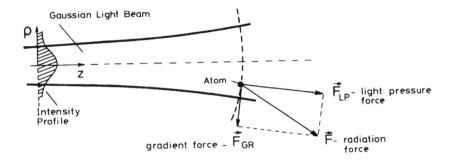

Fig. 5 Radiation pressure force acting on an atom in traveling Gaussian laser beam. The Gaussian laser beam exerts a gradient force on the dipole moment that intense light beam induces in atom. The resultant force of the laser field is not simply in the propagation direction of the laser beam.

$$\mathbf{F}_{RAD} = \mathbf{F}_{LP} + \mathbf{F}_{GR} \tag{2.15}$$

For a Gaussian laser beam, the light pressure force is a direct generalization of the force (2.10).

Depending on the detuning of laser radiation frequency with respect atomic transition frequency, the laser field either expels atom out of the beam ($\Omega > 0$) or draws it toward the beam center ($\Omega < 0$). In terms of atomic and laser parameter the gradient force is expressed as (Ashkin, 1978; Minogin and Letokhov, 1987):

$$\mathbf{F}_{GR} = \hbar(\rho/\rho_0^2)(\Omega - kv_z)[G(r)/(1 + G(r) + (\Omega - kv_z)^2/\gamma^2] \tag{2.16}$$

where the laser beam radius ρ_0, ρ the distance between the atom and the laser beam axis.

The dependence of the magnitudes and signs of the spontaneous and gradient forces are of entirely different character (Figs. 4 and 6). Unlike the light pressure force the value of gradient force is not limited as the saturation parameter is increased. The maximum value of the force increases proportionally to $G^{1/2}$. The velocity dependence of gradient force is determined by the curve of a dispersive form (Fig. 6).

The physical reason for the existence of the gradient force is an effect of a spatially non-uniform optical field on the atomic dipole moment which is induced by the laser field. From the standpoint of quantum mechanics, the gradient force results from the absorption and stimulated emission of photons by the atoms in a *non-uniform* laser field (Fig. 6a).

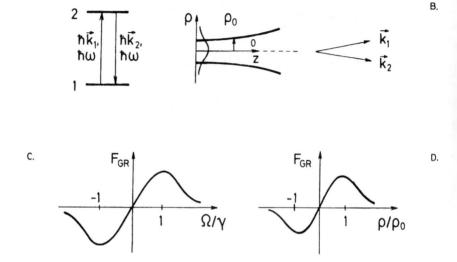

Fig. 6 Origin and properties of gradient (dipole), force F_{gr}: (A) absorption and stimulated emission of directed photons (k_1, ω_1), and (k_2, ω_2); (B) the Gaussian beam with transversal profiles of intensity I (ρ); (C) resonant dependence of F_{gr} as a function of frequency detuning Ω; (D) dependence of F_{gr} a a function of atomic displacement of atom ρ from optical axis.

Both forces (light pressure force and gradient) are essential to the control of atomic motion. For example, the gradient force allows one to trap atoms (channel them in one dimensional case) as illustrated in Fig. 3a. The spontaneous force makes it possible to cool atom and ions (Fig. 3b).

2.3 PLANE STANDING WAVE. TWO-LEVEL ATOM

This wave is formed from two plane waves traveling towards each other:

$$\mathbf{E} = \mathbf{e}E_0 \cos(kz - \omega t) + \mathbf{e}E_0 \cos(-kz - \omega t) = \mathbf{e}2E_0 \cos \omega t \cos kz \quad (2.17)$$

At low radiation intensity ($G \ll 1$) and for a two level atom the radiation force in the standing wave is determined by the sum of the forces (2.10) from each traveling waves.

However, at high saturation G, this no longer holds. The interaction between two-level atom and the two counterpropagating waves at $G > 1$ gives rise to the new effects beyond the scope of the spontaneous light pressure and gradient forces.

In order that we can consider classically the motion of the atom in such wave, it is necessary that the following condition be fulfilled:

$$p_z \gg \hbar k \qquad (2.18)$$

where $p_z = Mv_z$, the component of atomic momentum along the standing wave. It can be clearly seen that only in this case, according to uncertainty principle and with the reasonable relation $\Delta p_z < p_z$, do we uncertainty of coordinate $\Delta z < \lambda$. We assume also the following condition on the atomic and laser parameters:

$$kv_z < \gamma \qquad (2.19a)$$

$$kv_z < \Delta^2/(\gamma G_0^{1/2}) \qquad (2.19b)$$

where G_0 is the transition saturation parameter of one traveling wave. The inequality (2.19) reflects the absence of the Landau–Zener transition (Kazantsev, 1985). Also, we assume that the relation $t_{int} \gg \gamma^{-1}$.

2.3.1 Light pressure force

In a single standing light wave, two resonances occur under weak saturation, one for each traveling waves: $\pm kv_z = \Omega = \omega - \omega_0$ (Fig. 7). At $\omega < \omega_0$ the counter propagating waves decelerates the atom, whereas, its co-propagating counterpart accelerates it. The combine effect of the two traveling waves acting independently is described by a curve of dispersive character (Fig. 7). At exact resonant ($kv_z = 0$), the friction force is zero, and the slope of the curve at $v_z = 0$ gives the friction coefficient η for atom, which governs its cooling rate.

For low velocity of atoms:

$$kv \ll \gamma$$

the sum of the two gives a damping linear in the atomic velocity:

$$F_{LP} = \eta v \qquad (2.20)$$

$$\eta \cong -4\hbar k^2[(\Omega/\gamma)G/(1 + \Omega^2/\gamma^2)^2] \qquad (2.21)$$

where η is a damping (or friction) coefficient. In a weak standing wave at $\omega < \omega_0$, the traveling wave propagating counter to the moving atom cools it. But when saturation is strong, the stimulated re-emission of photon from one traveling wave into the other becomes predominant. At $\omega < \omega_0$ this stimulated re-emission change the sign of

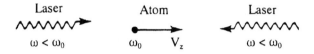

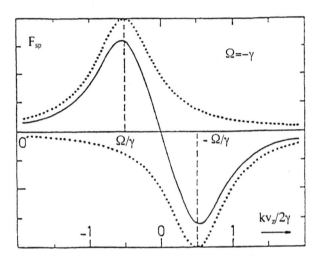

Fig. 7 Dependence of light pressure force as a function of velocity projection v_z for traveling wave (dotted line) and for standing wave (solid line); frequency detuning $\Omega=\omega-\omega_0=\gamma$.

friction coeffieient so that in the neighborhood of $kv \cong \gamma$ the cooling of the atom turns to its heating [Minogin and Serimaa, 1979]. However, where the frequency detuning of the laser field is positive ($\omega > \omega_0$), such re-emission of photons gives rise to a new cooling force that depends on stimulated emission and light induced shift of energy levels (see 2.3.3.).

2.3.2 Gradient force

A spatial modulation of the laser field intensity in the standing wave with a period $\lambda/2$ begins to show up with increasing radiation intensity, which leads to the appearance of the gradient force of potential nature:

$$\mathbf{F}_{GR} = -(dU_g/dz)\mathbf{e}_z \tag{2.22}$$

The associated potential has the form [Gordon and Ashkin, 1980]:

$$U_{GR} = (\hbar\Omega/2)\ln(1+s) \qquad (2.23a)$$

where

$$s = G/[1 + (\Omega/\gamma)^2] \qquad (2.23b)$$

is an off-resonance atomic transition saturation parameter, $G = 4G_0\cos^2(kx)$ is the atomic transition saturation parameter the standing light wave and $G_0 = d^2E^2/2\hbar^2\gamma^2$. In the limiting case of weak saturation and great frequency detuning,

$$G \ll [1 + (\Omega/\gamma)^2] \qquad (2.24)$$

the expression for the potential U_{GR} reduces to the simple form of the light wave, and a negative detuning ($\Omega < 0$), they coincide with the loops of the wave:

$$U_{GR} = (2\hbar^2\omega_R^2/\Omega)\cos^2(kx) \qquad (2.25)$$

where $\omega_R = dE_0/\hbar$ is the Rabi frequency. It follows from expressions (2.23) and (2.25) that the period of the potential is equal to the half the optical wavelength. At positive frequency detuning ($\Omega > 0$), the potential wells coincide with the nodes of the light wave, and at a negative detuning ($\Omega < 0$), they coincide with loops of the standing wave.

2.3.3 Stimulated (retarded) force

As noted in §2.3.1 the saturation parameter is increased, the force–versus–velocity curve starts developing the high-order resonances due to the nonlinear interaction between atom and the both counter-propagating waves. Stimulated transition in a standing light wave, which destroys the cooling effect in the case of strong saturation at a negative frequency detuning, can nevertheless be used to cool atoms, but at appositive detuning this time [Kasantsev, 1974; Kasantsev, 1985; Dalibard and Cohen-Tannoudji, 1985]. That this is possible was demonstrated with success in [Aspect et al., 1986].

To gain an insight into this interesting effect, account should be taken of two-quantum stimulated transitions in the standing light-wave field, which give rise to the spatially periodic atom-field interaction potential U(z) (or the gradient force F(z) = grad U(z)) in quantum mechanical terms, a light induced shift of atomic energy level that also oscillates in space. Combined with the spontaneous transition of the atom moving in the standing light wave, this give rise to the friction force [Kasantsev, 1974; Kasantsev et al., 1985; Dalibard and Cohen-Tannoudji, 1985].

The quantum mechanical treatment of the interaction of a two-level atom with a standing light wave shows also that the total radiative force depends on the atomic velocity. In a zeroth order of approximation of this force (for small parameter $(kv_z/\gamma) \ll 1$)) the gradient force (2.23) appears, and in first order — the friction force [Dalibard and Cohen-Tannoudji, 1985; Kasantev et al., 1986]

$$\mathbf{F}_f = 2\hbar(\Omega/\gamma)G\{[(1 + (\Omega/\gamma)^2 - G(1 + G/2)]/[(1 + (\Omega/\gamma)^2 + G)^3]\}*$$

$$\tan^2(\mathbf{kz})\mathbf{k}(\mathbf{vk}) \qquad\qquad (2.26)$$

which is called stimulated or retarded force.

Let us give the explanation of appearance of this force in two physically equivalent languages. The well-known light-induced shift of the energy levels of the two-level atom is given by [Cohen-Tannoudji, 1961; Carver, 1961; Mizushima, 1964; Bonch-Bruyevich et al., 1968]

$$d\omega \cong (1/4)E^2(d^2/\hbar^2\Omega) \qquad\qquad (2.27)$$

The levels are shifted one and the same distance but in opposite directions. With $\omega < \omega_0$, the perturbation caused by light field increases the distance between the level, and vice versa with $\omega > \omega_0$, it decreases the spacing between them (Fig. 8).

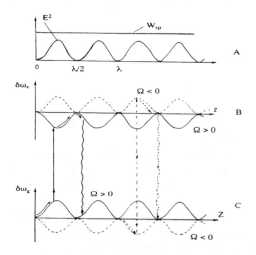

Fig. 8 (A) Distribution of standing wave intensity E^2 and the spontaneous and stimulated emission probabilities W_{sp} and W_{st}; (B) and (C) energy shift $\delta\omega_g$ of the ground and excited states for positive ($\Omega>0$) and negative ($\Omega<0$) light field frequency detuning. Vertical straight lines corresponds to induced absorption, vertical wavy lines — to spontaneous emission.

The stimulated cooling of the atoms in a standing light wave can occur on account of the following predominant sequence of stimulated and spontaneous processes. Since the rate of stimulated processes W_{st} (stimulated two-quantum processes responsible for the light-induced shift of energy levels and stimulated absorption responsible for excitation of atom) is proportional to E^2, in the standing light wave loops there predominantly occurs the excitation of the atom upon absorption of the red shifted light (Fig. 8). If the condition

$$kv_z \approx \gamma \qquad (2.28a)$$

is satisfied, the excited atom may enter the region of standing wave where the field intensity is its minimum and the light-induced level shift are small, and then spontaneously emit a shorter-wavelength photon. The spontanous emission rate is independent of the periodic field intensity variations. The energy difference between the absorbed and the spontaneously emitted photon is derived from kinetic energy of the atom. Such an absorption of photon in the standing light wave loops and their spontaneous re-emission at the field minima may take place repeatedly, thus gives rise to a friction force (2.26) proportional to the atomic velocity.

A clear and consistent theoretical description of the cooling proccess in a standing light wave was made in [Dalibard and Cohen-Tannoudji, 1985] on the basic of the dressed-atom approach [Cohen-Tannoudji and Reumond, 1978] providing a qualitative understanding of the main characteristic of the stimulated force (its mean value, fluctuations, velocity dependence, etc.) in high intensive limit.

An important feature of stimulated force in a standing light wave (which is a consequence of the light induced energy level shift) that it cannot be saturated by increasing the light field intensity [Kasantsev, 1974; Kasantsev et al., 1985; Dalibard and Cohen-Tannoudji, 1985, Aspect, 1988]. This is naturally explained by the fact that the atomic cooling rate is proportional to the energy difference between the absorbed and re-emited photons that is governed by the magnitude of light induced shift (2.27) proportional to the field intensity.

2.4 AN ARBITRARY MONOCHROMATIC WAVE. MULTI-LEVEL ATOM

When monochromatic wave interacts with multi-level atom, the number of new effects grows still greater. A general description of radiation forces with arbitrary number of degenerate or nearly degenerate sublevel and in an arbitrary monochromatic radiation field was given in [Nienhuis, 1993]. It allows also for an arbitrary magnetic field which adds Zeeman precession to atomic evolution. The total radiative force on the atom may be generally separated into four terms. Apart from the radiation pressure

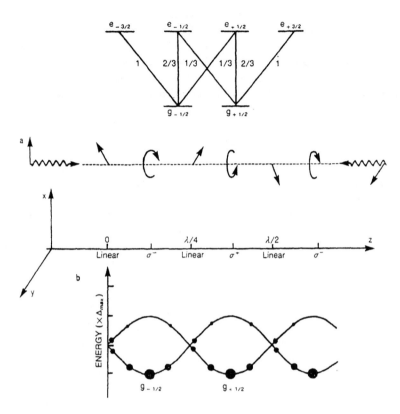

Fig. 9 Explanation of atomic cooling based on the effects of polarization gradient, light induced level shifts and optical pumping effect for multilevel atom (Cohen-Tannoudji and Phillips, 1990).

and the gradient force (which we have considered in 2.3.1 and 2.3.2) there are two terms that are from the gradient of the polarization direction. One of these term corresponds to fluorescent scattering of absorbed photons. The other term arises from the redistribution of photons between plane waves that compose the radiation field.

The experiment [Lett et al., 1988; Shevy et al., 1989; Dalibard et al., 1989; Sheehy et al., 1989; Weiss et al., 1989; Sheehy et al., 1990; Shang et al., 1990] have demonstrated that cooling below the Doppler limit can occur for the system having multistate structure in the field of driven transition. An essential ingredient for this damping force is a variation of the relative orientation of the atomic dipole with the respect to the light polarization during the transversal of a wavelength. Such situation arises when atom moves through a field with polarization gradient [Dalibard and Cohen-Tannoudji,

1989; Urgan et al., 1989] or when a magnetic field causes Zeeman precession of the atomic dipole [Sheehy et al., 1989; Weiss et al., 1989].

Let us consider briefly the new possibilities that are being opened up to here.

2.4.1 Polarization gradient force

The effect of periodic modulation of the light-induced shift of atomic energy level (see (2.27) and Fig. 8) is especially manifest in the case of three-level atom where it enables new mechanism of deep (sub Doppler) atomic cooling and trapping to come into play. An excellent description of the mechanism allowing the Doppler cooling limit to overcome been presented in the brief review [Cohen-Tannoudji and Phillips, 1990]. These mechanism include, in addition to the above mentioned effect of light-induced energy shifts, also those of optical pumping and laser polarization gradients [Dalibard et al., 1989; Chu et al., 1989].

The two enconter-propagating laser beams with orthogonal linear polarization form a standing light wave whose local polarization changes every eight's of a wavelength from linear to circular type (Fig. 9a). Depending on the sign of the local circular polarization, there takes place, thanks to the optical pumping, the accumulation of atoms either at the sublevel $g_{-1/2}$ (for σ^-) or at the sublevel $g_{+1/2}$ (for σ^+). Thus, the spatially periodic modulation of the circular polarization sign causes a spatially periodic variation of the population of the Zeeman sublevels of the ground state. The shift of Zeeman sublevels also depends on the circular polarization sign: the σ^+ wave shifts only the $g_{+1/2}$ sublevels, whereas its σ^- counterpart, only the $g_{-1/2}$ ones. So, the light shift energies and population of two ground state sublevels of the atom vary with the local polarization of the standing light wave, hence with the atom's position, as shown in Fig. 9c where the sublevel population are proportional to the size of the full circles.

Let the atom move at a small angle to the light wave front, so that its velocity projection is

$$0 < v_z \ll \gamma/k \qquad (2.28b)$$

i.e., much smaller than required by condition (2.28a). Consider for the case of definiteness an atom starting to move from the point $z = \lambda/8$ (Fig. 9). Such an atom will cover the distance of $\lambda/4$ in a time shorter than optical pumping time τ_p, and so it will climb from the potential minimum to the maximum, while residing at one and the same sublevel, as illustrated in Fig. 9. The probability that the optical pumping will cause the atom to move from one sublevel to the other here becomes higher. But the spontaneous decay time is much shorter than the optical pumping time, $1/\gamma \ll \tau_p$, and therefore the spontaneous transition of the atom to the other sublevel takes place while it travels the distance of $\Delta z \ll \lambda/4$. As a result the atom absorbs long-wave photon and emit short wave ones. The energy difference between these photons is equal to the level shift (2.27), and so it is proportional to the light field intensity. Thus, the rate of atomic cooling by this mechanism is proportional to the laser intensity. Naturally this process of absorption at the top of potential hill and spontaneous descent in the valley takes place repeatedly, providing for the deep cooling of the atom to that of the order of recoil velocity v_{rec}. A detailed theoretical analysis of this atomic cooling mechanism has been made in [Dalibard and Cohen-Tannoudji, 1989; Urgan et al., 1989].

Its substantial difference from the Doppler mechanism is that it operates effectively at atomic velocities too low for Doppler mechanism to be efficient and makes it possible to reach temperatures (velocities) much below the Doppler limit (2.13), down to the recoil velocity and temperature (Dalibard and Cohen-Tannoudji, 1989):

$$v_{rec} = \hbar k/M \quad \text{or} \quad T_{rec} = R/k_B \qquad (2.29)$$

where R is the recoil energy.

2.4.2 Magneto-optical forces

In the presence of static magnetic field the radiative forces acting on a multilevel atom in a laser field may acquire new properties [Raab et al., 1987; Sheehy et al., 1990]. Of particular interest for the creation of new types of atomic traps is a confining potential — like character of the radiation force. To produce a magneto-optical force pseudopotential character [Raab et al., 1987], a scheme based on the spontaneous force has been used. In this case, the spatial dependence of this force is determined by the Zeeman shift in the magnetic sublevel of the atom induced by spatially-nonuniform static magnetic field. One of the short comings of this scheme is that the maximum of this magneto-optical force is restricted by the upper limit of spontaneous force $\hbar k \gamma$, where 2γ is the natural width of exited state.

It has been shown [Grimm et al., 1992] that there is magneto-optical force based on the induced photon re-emission between two counter propagating light waves having different direction of polarization vectors. It is essential that the value of this magneto-optical force may exceed $\hbar k \gamma$.

2.5 VELOCITY-SELECTIVE COHERENT POPULATION TRAPPING

Until recently the control of atomic motion has been treated in the semiclassical range of parameters where the wave characters of atomic translation is inessential (True, some questions of the quantum atomic motion in a standing laser light wave were treated theoretically (Letokhov and Minogin, 1978).

The classical character of atomic motion allows for the application of the kinetic equation approximation that has long been used for atomic motion control problems. It makes it possible to describe the evolution of a atomic ensemble in terms of physical kinetics and use the notion of light pressure force, the coefficient of atomic momentum diffusion in a radiative field, the atomic ensemble temperature, and so on. The kinetic equation approximation holds true if the characteristic atomic momentum distribution width exceeds the magnitude of the momentum imparted to the atom by photon in a single absorption or emission event. In the semiclassical approximation, the minimum cooling temperature is determined by the recoil effect (2.29).

The investigation conducted by Cohen-Tannoudji and co-workers have demonstrated that the fundamental atomic cooling limit defined by recoil energy (2.29) can be also overcome in a three-level-scheme by means of a new mechanism based on the *velocity selective coherent population trapping* effect [Aspect et al., 1988]. This effect consists in the absence of absorption of light by an atom in two coherent light waves when the atom is in a superposition state at two sublevel of the ground state. This is only possible for stationary atoms, and so there occurs the accumulation (and not cooling) of atom in the superposition state with their velocities lower than the recoil velocity v_{rec}. The description of the process must be entirely quantum-mechanical.

Let us give first the notion of coherent population trapping. Consider a quantum system characterized by nonperturbed Hamiltnian H_0 in an external field, the interaction between the system and the field being described by the perturbation operator $V(t)$. The temporal evolution of the state of the system in the interaction representation is described by the Schrödinger equation

$$i\hbar(d/dt)/\Psi> = \exp(i\hbar^{-1}H_0t)V(t)\exp(-i\hbar^{-1}H_0t)/\Psi> \qquad (2.30)$$

In some cases, the Schrödinger equation can have very special solution $/\Psi^{n.a.}>$ satisfying the equation

$$\exp(i\hbar^{-1}H_0t)V(t)\exp(-i\hbar^{-1}H_0t)/\Psi^{n.a.}> = 0 \qquad (2.31)$$

Equation (2.31) means that at any moment of the time there is no interaction between the system and the external field. Such solution possess some extraordinary properties. Let the state $/\Psi^{n.a.}>$ (the coherent population trapping state, the CPT-state,) at the initial moment of time be composed of a certain finite set of the unperturbed Hamiltonian eigenfunction $/\Psi_n>$. At any subsequent moment, the state of the system will be made up of the same states $/\Psi_n>$ despite their extensive intermixing under the effect of the perturbation $V(t)$. The system is as if "trapped" in these states. This phenomenon has come to be known as the *coherent population trapping (CPT)*[Arimondo, 1992]. If, in addition, the CPT state is composed of the ground states ot the system only, the allowance made for their interaction with electromagnetic vacuum field leads to the accumulation of atom in the CPT-state

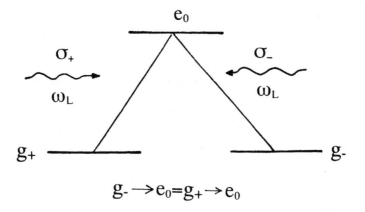

Fig. 10 Λ — configuration of atom for which coherent population trapping takes place.

on account spontanoues decay of the upper levels. The CPT-states play the part of attracting multitudes in the state space of the given quantum system.

In the work reported in [Aspect, 1989], they treated the problem of CPT in a system consisting of a three-level-atom and an electro-magnetic field of $\sigma^+\sigma^-$ configuration (Fig. 10). The lower states of the atom taken to be the the m $=$ ± 1 components of Zeeman triplet, and the upper state, the $m' = 0$ component of some other triplet. The translational motion of the atom was described in quantum-mechanical terms. In such a field the atom has nonabsorhing state composed of two lower states with m $=$ ± 1 and $p_z = \pm\hbar k$ (The CPT-states of such atom turned out to be localized in two parallel planes $p_z = \pm\hbar k$ in the momentum space, OZ being the field propagation direction).

The CPT-states exist only for atoms with $v_z = \pm \hbar k/M$. For another v_z, the interference between the two amplitude of the probability of transition from the lower sublevel to the the excited states ceases to be destructive. As a result, the atoms at any one of the sublevels can absorb a photon and thus give to excited state. This circumstance was reflected in the name of the phenomenon — *velocity selective coherent population trapping (VSCPT)*.

Being subject to spontaneous decays, atoms perform Brownian motion in the momentum space. Should the atom, while wandering, find itself in the CPT-state, it remains in this state at every subsequent moment of time. In the course of the time, practically all atoms accumulate in the CPT-state and become localized in the $p_z = \pm \hbar k$ planes in the momentum space. Such a narrowing atomic velocity distribution can be interpreted as one-dimensional ultra-deep cooling [Aspect et al. 1988]. What is important is that this cooling mechanism is based not on the use of the friction force, but on the accumulation of atoms in these nonabsorbing CPT-states.

Recently a generalization of velocity selective coherent population trapping to two dimensional and three dimensional was also proposed [Mauri and Arimondo, 1991; Ol'shani and Minogin, 1991].

2.6 ATOMIC MOMENTUM DIFFUSION

Let us consider now the fluctuation processes which cause a diffusion of the atomic momentum against the background of its drifts under the action of the radiation force [Pussen, 1976]. There are two such fluctuation processes. One process consists of a fluctuation in the *direction* of the spontaneous emissions of photons. Since the change in the momentum of an atom $\delta\mathbf{p}$ is related to the momentum of the spontaneously emitted photons:

$$\delta\mathbf{p} \equiv \mathbf{p}' - \mathbf{p} = -\hbar\mathbf{k}_s \qquad (2.32)$$

the fluctuation in the direction of spontaneous emissions always lead to fluctuations in the directions of the recoil momentum $\hbar\mathbf{k}_s$.

The second fluctuation process is caused by fluctuations in the *number* of photons which are scattered by atoms. Since a recoil momentum $\hbar\mathbf{k}$ is associated with each stimulated absorption (or emission) of a photon, the momentum of the atoms fluctuates by the amount $\pm\hbar\mathbf{k}$ upon each unit fluctuational change in the number of scattered photons. This fluctuation process causes the momentum of the atoms to vary along the wave vector $\mathbf{k}$, always by a discrete quantity $\hbar\omega/c$.

The fluctuational deviation of the momentum of the atom from its expectation value is:

$$\delta\mathbf{p} = \mathbf{p} - \langle p \rangle = (\mathbf{p}_0 - \langle p_0 \rangle) + \hbar\mathbf{k}\,\delta N_i + \Sigma\hbar\mathbf{k}_s \qquad (2.33)$$

Here $\delta N_i = N_i - \langle N_i \rangle$ is the deviation of the number of photons scattered by the atoms from the expectation value. The second term in (2.33) takes into account fluctuations in the number of scattered photons, while the third term takes into account fluctuations in the direction of the spontaneous emission of photons. These fluctuation processes cause a diffusive broadening of the momentum distribution of atoms. Each of the fluctuation processes causes a corresponding type of diffusive broadening of the momentum distribution. The fluctuations in the number of scattered photons are the reason for the *directed diffusion*, while a fluctuation in the direction of spontaneous photons are the reason for *anisotropic diffusion*. These diffusion processes are determined quantitatively by corresponding diffusion tensors.

$$D_{ij} = 1/2\langle(\delta p_i \delta p_j)\rangle/\delta t \qquad (2.34a)$$

where δp_i is the deviation of momentum projection p_i from their average value $\langle Pi \rangle$.

For the simplest case of the interaction of a two level atom with resonant radiation with wave vector k and a frequency $\omega = kc$, we can evaluate the diagonal elements of the momentum diffusion tensor in a simple way. We define the diagonal elements of the diffusion tensor by

$$D_{ii} = 1/2 \langle (\delta p_i)^2 \rangle / \delta t \qquad (2.34b)$$

where $i = x, y, z$. If the radiation is propagating along the z-axis, we can write on the basis of (2.34a) the following expression for the diffusion tensor:

$$D_{ii} = D_{ii}^a + D_{ii}^d = 1/2 \hbar^2 k^2 [\alpha_{ii} \langle N_s \rangle / \delta t) +$$

$$+ (\langle (\delta N)^2 \rangle / \delta t) \delta_{zi}] \qquad (2.35)$$

here $\alpha_{ii} = \langle \cos^2 \Theta_i \rangle$ is the expectation value of the square of cosine of the angle Θ_i, which determines the projection of the momentum $\hbar k$ onto the axis $i = x, y, z$, and $\langle N_s \rangle$ is the average number of reemitted photons. The indices "a" and "d" specify the an anisotropic and directed diffusion.

To find the elements of the momentum diffusion tensor we need to determine the mean square value of the fluctuations in the number of photons which are involved in the stimulated scattering $\langle (\delta N)^2 \rangle$.

For a two level atom the atomic momentum diffusion in a traveling wave [Baklanov and Dubezkii, 1976; Pussep, 1976] is

$$D_{ii} = \hbar k \gamma (\alpha_{ii} + \delta_{zi}) \left[G / (1 + G + (\Omega - kv)^2 / \gamma^2) \right] \qquad (2.36)$$

The diffusion tensor has the Lorentzian dependence on the velocity of atoms and reaches its maximum value at exact resonance of atomic transition with laser frequency.

The atomic momentum diffusion tensor in a standing wave for a two level atom is a very complicated function of the atomic center of mass coordinates, the velocity of the atoms, the saturation parameter and the detuning of the laser frequency from an absorption frequency. In the important case of small velocity the diffusion coefficients can be expressed as a sum of two diffusion coefficient as in the case of a traveling wave [Gordon and Ashkin 1980; Minogin and Letokhov, 1987]:

$$D_{sw} = D_1 + D_2 \qquad (2.37)$$

where

$$2D_1 = \hbar^2 \alpha^2 \gamma [s/(1+s)^3]\{1 + [4\gamma^2/(\gamma^2 + \Omega^2) - 1]s + 3s^2 +$$

$$+ [(\gamma^2 + \Omega^2)/\gamma^2]s^3\}\{1/[1 + A(\alpha v/\gamma)^2]\} \qquad (2.38)$$

is the directional diffusion coefficient, and

$$2D_2 = \hbar^2 k^2 \gamma [s/(1+s)]\{1 + A(\alpha v/\gamma)^2]\} \qquad (2.39)$$

is the isotropic diffusion coeffient. In (2.38) and (2.39) $s = 4s_0 \cos 2kx$ is a saturation parameter in a standing wave and s_0 – in one of the traveling wave:

$$s_0 = (\omega_{0R}^2/2)/(\gamma^2 + \Omega^2) \qquad (2.39a)$$

and $\alpha = \text{grad ln u}$, u is the local field amplitude.

The expression for the diffusion coefficient less than factor inside the second set of braces in (2.38) and (2.39) is the first term of the diffusion coefficient expansion in term of velocity. For small velocities (less than critical velocity the coefficient A approximately equal: $A \cong 3$ [Balykin et al., 1989a].

For the case of weak saturation of the atomic transition and zero velocity the total diffusion coefficient in standing wave is

$$2D = 4\hbar^2 k^2 \gamma s_0 \qquad (2.40)$$

In the case of large saturation intensity ($s \gg 1$) and zero velocity the diffusion coefficient equals:

$$2D = \hbar^2 k^2 \gamma ((1 + 16G_0 \sin^2 kx) \qquad (2.41)$$

where $G_0 = I_0/I_S$.

From (2.40) and (2.41) we can see that in the case of a small saturation intensity diffusion coefficient does not depend on the spatial coordinate, in the case of large saturation intensity we have the same spatial dependence of the diffusion coefficient as for laser intensity in a standing wave.

3. COLLIMATION OF ATOMIC BEAM

The need to collimate a beam of particles arises whenever one has 1) to increase their phase space density (the goal is to compress a spatial and a velocity distributions of the atoms) and 2) to increase a spatial coherence of atomic beam. The method of collimation will depend on the particular particles species. Common for all these methods is to use dissipative processes, because the Liouville theorem tell us that conservative forces cannot alter phase space density. For example, light charge particles are collimated by radiative friction; protons, antiprotons and ions — by electron cooling; the heavy particles — by ionization loss.

The dissipative force most effective for neutral atomic beam is the laser radiation pressure. The atomic beam collimation experiments performed up to date have used the radiation pressure force [Balykin et al., 1984; Balykin and Sidorov, 1987a], the stimulated (retarding) force [Salomon et al., 1987], the polarization gradient force [Dalibard et al., 1989; Chu et al., 1989], the magnetic-field – induced molasses [Sheery et al., 1990] and the coherent population trapping [Aspect et al., 1986]. Each of these schemes has its own advantages and disadvantages. The radiation pressure force can be used to obtain considerably narrow angular divergence of atomic beam. A preference of using the retarding force is a faster beam collimation than in the first case however the ultimate collimation angle achievable by this means is much larger. The collimation by the polarization gradient force permits to reach the very small divergence of a beam and a characteristic time for collimation is also short but a capture range in this method is also very small. The coherent population trapping method has no limit for a final divergence of atomic beam but it demands considerable time for collimation and only a small fraction of atoms, whose velocity originally inside of the capture range, will be in a final narrow collimated atomic beam.

The method of increasing the spatial coherence by using special scheme of collimation is only one so far known for atomic beam. (In optics of photons the high spatial coherence of a light beam can be obtained in a laser through a process of a stimulated emission).

3.1 COLLIMATION BY LIGHT PRESSURE FORCE

Let us first consider the collimation of an atomic beam by the light pressure force in more detail. Figure 11 shows one of the possible scheme for the collimation (a transverse radiation cooling) of an atomic beam. In this arrangement, the beam of atoms emerging from the source is irradiated by axisymmetric light field, whose frequency ω is red shifted with respect to the atomic transition frequency ω_0. For two dimensional collimation of the atomic beam the axisymmetrical field is formed by

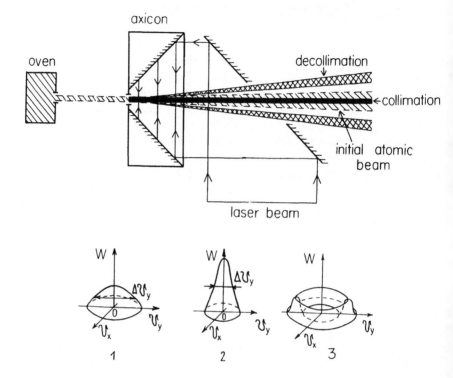

Fig. 11 Collimation of atomic beam by means of laser light. The atomic beam is passing through a conical mirror reflector and irradiated from all sides by laser light [Balykin et al., 1984].

the reflection of a plane light wave from a conical mirror surface (reflection axicon). This axisymmetrical field permits two dimensional collimation of an atomic beam. In this field an atom whose velocity is directed from the axis of the cone experiences a radiation pressure force. This force is directed opposite to the radial velocity vector $\mathbf{v}_\rho$ if $\omega < \omega_0$ and along $\mathbf{v}_\rho$ if $\omega > \omega_0$.

The reason for this is the radial direction of wave vector $\mathbf{k}$ of the field inside of axicon. An atom moving at an angle to the axis of the cone interacts in the XY plane with two counter propagating light waves whose intensities are equal at any point in space. In the rest frame of the atom one of the waves has a frequency $\omega - kv_\rho$, while the other has a frequency $\omega + kv_\rho$. When the laser frequency is chosen lower than the frequency of the atomic transition, the atom absorbs more efficiently the photons from the wave that is propagating counter the radial velocity. This means that the force acting on the atoms for $\omega < \omega_0$ is directed counter the radial velocity.

The force in the inner region of the axicon for $\omega < \omega_0$ causes a rapid narrowing of the transverse velocity distribution of the atomic beam, leading to a decrease in the angular divergence of the beam and an increase in the density of atoms.

Let us make some simple estimation of the degree of collimation of an atomic beam which is moving at the thermal velocity at room temperature in a transverse standing laser field. The evolution of the transverse atomic velocities is determined both by radiation pressure, which is narrowing the transverse velocity distribution, and the diffusion of atomic moment, which broadens the distribution. For transverse velocity $v_\perp$ satisfying the condition $v_\perp < \gamma/k$ (the transverse velocity of atoms is less than a capture range the light pressure force) and $G \leq 1$, the radiation pressure force reduces to a friction force:

$$F = -\eta v \qquad (3.1)$$

where the dynamic friction coefficient η is given by

$$\eta \cong 4(\hbar k^2)(\Omega/\gamma)G[(\gamma^2 + \Omega^2)/\gamma^2]^{-2} \qquad (3.2)$$

Here k is the wave vector, 2γ is the natural line width, G is the saturation parameter of atomic transition, Ω is the detuning of the laser frequency relative to the atomic transition frequency, M is the mass the atom.

The equation of motion of an atom under the influence of the friction force (3.1) implies that the characteristic time for narrowing of the velocity distribution is determined by the value of $t_c = \eta^{-1}$. For the typical laser parameters $\Omega/\gamma = 2$, $G = 1$, the characteristic collimation time is $t_c = 2 * 10^{-5}$ sec. If the time required for atoms to transverse the laser field is greater than the time t_c, the evolution of the atomic ensemble in the laser field is determined not only by radiation pressure but also by diffusion of the atomic momentum.

The combined effect of radiation pressure and momentum diffusion leads to the establishment of a steady velocity distribution of the atoms, with an effective temperature [Letokhov et al., 1977; Wineland and Itano, 1979]:

$$T_{min} = (\hbar\gamma/2k_B)(\gamma/\Omega + \Omega/\gamma) \qquad (3.3)$$

where k_B is the Bolzman's constant. Using this expression of the temperature, one can find a collimation angle of the atomic beam at the exit of the laser field:

$$\delta\varphi_{min} = (1/\langle v \rangle M)^{1/2}(2k_B T_{min})^{1/2} = (1/\langle v \rangle)(\hbar\gamma/M)^{1/2} \qquad (3.4)$$

Here $\langle v \rangle$ is the average velocity of atoms along the axicon. For a thermal atomic beam the collimation angle is of the order of $10^{-3} - 10^{-4}$ rad. The use of collimation can lead to an increase in the density of atoms: $\rho_{int}/\rho_{fin} = (\delta\varphi_{in}/\delta\varphi_{fin})^2$. The increase in the phase density can reach a value of $(\delta\varphi_{in}/\delta\varphi_{fin})^4$. The initial divergence of atomic beam $\delta\varphi_{in}$ is determined by the capture range ($v_c \approx \gamma/k$) of the force and a longitudinal velocity $\langle v \rangle$: $\delta\varphi_{in} = v_c/\langle v \rangle \approx \gamma/k\langle v \rangle >$; the final divergence of a beam is determined by the expression (3.4). In the case of collimation by radiation force the increase of density can be:

$$\rho_{int}/\rho_{fin} \approx v_c/v_r \qquad (3.5)$$

where $v_r = \hbar k/M$ is the recoil velocity.

The capture range of collimation by a plane laser field can be increased from $v_c \approx \gamma/k$ to a large value by using a expanded convergent laser field [Nellessen et al., 1989; Aspect et al., 1990]. The atoms moves in such field approximately along the wavefront and its velocity is perpendicular to wave vector of the field. Instead of expanding the laser field it is possible also to use a pair of a concave and a convex mirrors having a common center of radius [Shimizu et al., 1990]. The atomic beam was simultaneously deflected and collimated in this last scheme. The observed intensity gain was about 30.

As an example let us consider the collimation of a sodium atomic beam. Figure 11 shows the experimental set up for atomic beam collimation [Balykin et al., 1984; Balykin and Sidorov, 1987a], which consists of the atomic oven, the conical reflection axicon and laser beam. The atomic beam divergence was originally $\delta\varphi_{in} = 3.2 * 10^{-2}$ rad. The length of interaction between atoms and laser field was 35 cm. In order to avoid an optical pumping process the four-level scheme of excitation was used in the experiment (Fig. 12). The two mode radiation was tuned to the D_2-line of sodium atoms. One frequency excited the atoms from the level $F = 1(3S_{1/2})$ to $F' = 2(3P_{3/2})$, the other one excited atom from $F = 2(3S_{1/2})$ to $F' = 3(3P_{3/2})$.

The atomic beam profile was measured by recording the fluorescence excited by an additional single-mode dye laser. For this purpose the laser radiation focused by a long-focus lens crossed the atomic beam in the plane of Fig. 11 at small angle. In the perpendicular plane the laser beam moved parallel to itself within several diameters of the atomic beam. The use of single mode radiation and an almost longitudinal intersection of the atomic beam with the laser beam allowed recording the atomic beam profile with definite longitudinal velocity.

Fig. 13 shows the profiles of the atomic beam before and after its interaction with the laser field for different laser detunings. There is optimal laser detuning when the atomic beam collimation is maximum. As the detuning is decreased, the degree of collimation drops drastically. Near zero detuning the broadening of the atomic beam can be clearly seen, and its intensity is decreased. This atomic beam broadening was caused by velocity diffusion of atoms. At positive detuning the atomic beam becomes essentially decollimated. The strongest decollimation is observed when its value is the same but the sign is opposite to the detuning of the optimal collimation. The atomic

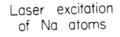

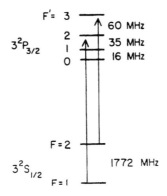

Fig. 12 Excitation of sodium atoms by two-frequency laser radiation for laser collimation.

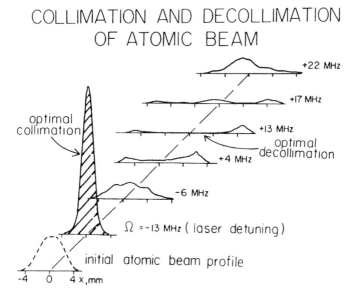

Fig. 13 Collimation of sodium atomic beam by light pressure force. Series of atomic beam profiles using different detuning of laser frequency.

beam became ring-shaped. It can be seen clearly from Fig. 13 that the collimation can be observed when the detuning ranges from the -30 to -8 MHz and decollimation occurs with the detuning between 0 and $+25$ MHz. Thus the strong influence of laser light on the atomic beam takes place in the frequency range of 55 MHz. The atomic beam intensity at the center of the beam was changed by more than 800 times.

3.2 COLLIMATION BY STIMULATED (RETARDED) FORCE

The stimulated force (or retarded force, §2.3.3.) is a velocity dependent force and it produces a heating of the atoms for a *negative* detuning and a cooling for a *positive* detuning. The retarded force is found to be maximum for velocities $v \leq \gamma/k$ and the capture range does not depend on an intensity light field. In a high intensive standing wave ($G \gg 1$) the retarded force is larger than the light pressure force and can be much more effective for a laser cooling and collimation of an atomic beam than the light pressure force. For a high intensity ($G \gg 1$) and the small atomic velocities ($kv \ll \gamma$) this force takes the following form (Dalibard and Cohen-Tannoudji, 1985):

$$F_{RT} = -(M/\tau)v = -\eta_{RT}v \qquad (3.6)$$

where τ is given by

$$1/\tau = (1/\tau_0)(\Omega/\Gamma)\left[(3/4)(1+S)^{1/2} + (3/2)/(1+S)^{1/2} - (1/4)/(1+S)^{3/2} - 2\right] \qquad (3.7)$$

and $\tau_0 = M/\hbar k^2$, $S = \omega_R^2/2\Omega^2$, ω_R — the Rabi frequency of one travelling wave, Ω — a laser detuning and $\Gamma = 2\gamma$. The friction coefficient is maximum at the laser detuning $\Omega \cong \omega_R/4$ and equals:

$$\eta_{RT} = (\hbar k^2/2)(2\Omega/\Gamma) \qquad (3.8)$$

The characteristic time of an atomic collimation by retarding force τ_{RT} is determined (as in the case of collimation by light pressure force) by friction coefficient and it equals

$$\tau_{RT} = \eta_{RT}^{-1} = \tau/(2G)^{1/2} \qquad (3.9)$$

where τ — the collimation time by the light pressure force. It is clear from (3.9) that for $G \gg 1$ a collimation by the retarding force can be faster than one by the light pressure force.

The combined effect of collimation by the force and the heating by the momentum diffusion leads to minimum effective transverse temperature of atomic beam T'_{min} and to corresponding minimum angular divergence $\delta\varphi'_{min}$:

$$T'_{min} \cong (\hbar\gamma/2k_B)(G/2)^{1/2} \qquad (3.10)$$

$$\delta\varphi'_{min} \cong (1/\langle v\rangle M)^{1/2} (2k_B T_{min})^{1/2} = (1/\langle v\rangle)(\hbar\gamma/M)^{1/2}(2G)^{1/2} \qquad (3.11)$$

The collimation of atomic beam by the retarding force was first observed with a cesium [Aspect et al., 1986] than with sodium atoms [Wang et al., 1989; Wang et al., 1990]. The cesium atomic beam [Aspect et al., 1986] with initial divergence 8 mrad was irradiated at right angle by intense one-dimensional standing wave. At the entrance of the interaction region, the initial velocity spread was 2 m/s. The final spatial profile was analyzed by a tungsten hot-wire detector. Before entering the interaction region, all atoms were optically pumped to one magnetic sublevel of ground state, which achieves a two-level system. The experimental results for the case of incident laser power 70 mW (the corresponding Rabi frequency of about 50 γ), the optimal detuning +6 γ and the transition time of interaction 15 μs are shown in Fig. 14. The atomic beam was strongly collimated to a narrow velocity peak of 40 cm/s (curve b) which is 5 times narrower than that of initial one (curve a). For the negative detuning ($\Omega = -6\gamma$) the atomic beam was decollimated and the atomic beam profile exhibits double-peak structure (curve c).

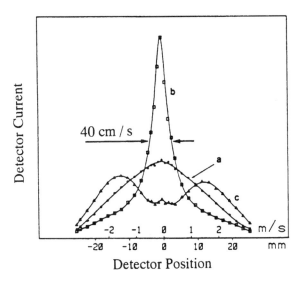

Fig. 14 Collimation of cesium atomic beam by stimulated (retarded) force. The initial spatial profile of atomic beam (curve a), after collimation (curve b) and decollimation (curve c) [Aspect et al., 1986].

Besides the many attractive feathers of the retarding force it was done only a few experiments with the use of this force. The main limiting factor is a radiation power from a laser which still is not sufficiently high.

At the end of this paragraph we will mention also the laser cooling experiments with the use of the retarding force [Prentiss et al., 1989; Begilow and Prentiss, 1990; Tollett et al., 1990].

3.3 COLLIMATION OF ATOMIC BEAM BY POLARIZATION GRADIENT FORCE AND IN A MAGNETIC FIELD

We have seen in (subsection 2.1.4.) that in a standing wave with a crossed or a circularly polarized light beams a multilevel atom is act upon a *polarized gradient force* (PGF). This force can be a friction force and for a small velocity. The friction coefficient can be larger than the friction coefficient of light pressure force. Later (Sheery et al., 1990) it was discovered that a similar force appears in a standing wave with an uniform polarization and in a magnetic field. In the first case the polarization gradient force comes from the spatial modulation an energy of ground state by light shift and from population transverses by optical pumping (2.4.1.).

The laser cooling of atoms by PGF is called now in literature by *optical molasses*. In the second case (called by a *magnetic-field-induced molasses*) the population transverses is a result of competition between optical pumping and Larmor precession [Sheery et al., 1990]. In a magnetic-field induced molasses the cooling depends on not only a sigh of frequency detuning but also on the atomic transition: for the transition $j \leftrightarrow j + 1$ the cooling can be obtained with red detuning; the transition $j + 1 \leftrightarrow j$ requires a blue detuning [Valentin et al., 1992]. The lowest achievable temperature in laser cooling by PGF in both cases is considerably smaller than the minimum temperature in the Doppler cooling but still remains larger than the recoil energy. It means that a minimum achievable divergence of atomic beam in a collimation by PGF will be also smaller then in the case of collimation by light pressure force and will be in within the angle range:

$$v_r/\langle v \rangle \; < \; \delta\phi_{min} \; \ll \; v_D/\langle v \rangle \qquad (3.12)$$

where $v_D = (\hbar\gamma/M)^{1/2}$. The capture range of polarization gradient force is also smaller than one for light pressure force.

The collimation of atomic beams by polarization gradient force was demonstrated as in one-dimensional and two-dimensional cases. In one dimensional collimation of Rb atoms by diode laser the minimum transverse temperature was 10 μK [Sheery et al., 1989] which is below the Doppler limit. In two dimensional collimation of Rb the brightness was increased by a factor more of than 20. The same group [Sheery et al., 1990] demonstrated the collimation of Rb beam by using a magnetic-field-induced laser cooling. In their scheme the atomic transverse velocity was damped by a standing wave of a circular polarized light in magnetic field of 20 μT. The transverse velocity was decreased to 2 cm/s which was also as in the previous case smaller than the Doppler limit (v_D = 10 cm/s).

3.4 COLLIMATION BY VELOCITY-SELECTIVE COHERENT POPULATION TRAPPING

In all previous collimation scheme, the minimum divergence of atomic beams was determined by minimum transverse velocity of atoms. The fundamental limit on the minimum velocity in laser cooling techniques is determined by recoil velocity. The group at the Ecole Normale Supérieure proposed a scheme of cooling of atoms by laser field in which a minimum velocity of atoms can be below recoil velocity. The mechanism of their cooling scheme is based on the effect of *the velocity – selective population trapping* (VSPT) (see 2.4.3.). The physical mechanism of the velocity – selective population trapping is the following. A 3-level V-shaped atomic system, through interaction with a laser field, is prepared in coherent superposition of two ground states g_+ and g_- different by momentum $\pm hk$ along the laser beam. Atoms in this state cannot more interact with laser light. The atomic momentum distribution along the laser axis exhibits a double peak structure. A mechanism of population of this non-absorbing state is an absorption-emission (spontaneous) cycles: through this cycles atom has a certain (usually very small) probability to be pump in this non absorption state. After sufficiently large number of such cycles considerable part of atoms can be accumulated in this state. The resulting transverse momentum distribution will consists of two peaks separated by δp = $2\hbar k$. The width of these peaks will be determined only by a time of interaction of atoms with a light and can be in principle arbitrarily small.

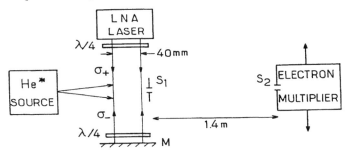

Fig. 15 Collimation of metastable He* atoms by velocity coherent population trapping [Aspect et al., 1988].

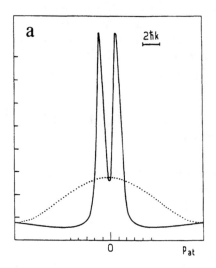

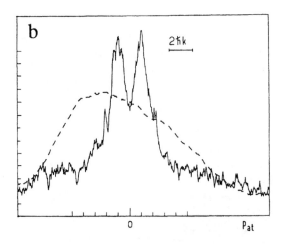

Fig. 16 Distribution of atomic momenta in the direction of laser beam for He* atoms after collimation by the velocity coherent population trapping [Aspect et al., 1988].

The one-dimensional collimation of metastable He* atoms by methods of velocity-selective coherent population trapping was demonstrated by Paris group (Aspect et al., 1988). Fig. 15 shows schematically their experimental set up. The atomic source at 77 K produces a monochromatic beam of metastable helium atoms. This atoms interacts with two σ^+ and σ^- polarized counter propagating laser light waves. There were stringent requirement for this experiment. The Zeeman sublevels g_+ and g_- must remain degenerate in the whole interaction region. This condition was fulfilled by compensation of magnetic field to 1 mG. The relative phase between both laser field must remain constant also in the whole interaction region. The overlap of the two beams was adjusted to 10^{-5} rad. In the experiment was measured the spatial transverse distributions of atomic beam from which the transverse velocity distribution deduced. Fig. 16 shows the transverses velocity distributions of atomic beam with and without collimating laser field. The momentum distribution exhibit double-peak structure, as predicted theoretically. The momentum width of each peak is smaller than the photon momentum.

Let us at the end of this chapter mention about some possible application of the collimation of an atomic beam. First, the methods make possible to change the atomic beam intensity more than one thousand times; second, it is possible to act on the velocity and spatial characteristics of an atomic beam with high frequency selectivity; third, to increase greatly a selectivity of interaction atomic beams with a hyperfine splitting of the ground state [Balykin and Sidorov, 1987a] and, fourth, to create the atomic beam with a high spatial coherence [Aspect et al., 1988].

4. FOCUSING OF ATOMIC BEAMS

4.1 Basic Idea and Different Approaches

The principal element in any sort of optics is a lens. It is therefore essential to create laser field configurations capable of focusing neutral atomic beams. There are at present two possibilities for focusing an atomic beam by laser light: by using the gradient force or by using the light pressure force.

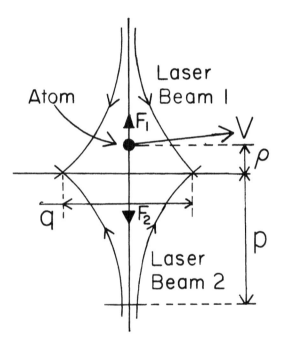

Fig. 17 Principle of focusing of an atomic beam with a laser lens: atomic lens formed by two Gaussian laser beams propagating in opposite direction (the x axis); on the x-axis the force is proportional to the atomic displacement and is directed towards the z axis.

Let us first find the condition that the radiation force meets in a laser lens. The lenses used in light and electron optics [Grivet, 1972] satisfy the following condition: a divergent concentric beam is transformed by means of the lens into a convergent concentric beam. This requirement means that the deviation angle $\delta\phi$ of an atom from its initial propagation direction in the paraxial optics approximation should be proportional to its displacement δr from the beam axis, i.e. $\delta\phi = -\alpha\delta\rho$, where α is the constant of proportionality. For this reason, the change in the transverse atomic velocity component will be $\delta v_\perp = -\delta\phi v_{\text{long}} = -\alpha\delta\rho v_{\text{long}}$, where v_{long} is the longitudinal atomic velocity. On the other hand, the change of the transverse atomic velocity is

$$\delta v_\perp = Ft_{\text{int}}/M = FL/Mv_{\text{long}} \qquad (4.1)$$

where F is the force exerted on atoms, M is the atomic mass, t_{int} the interaction time and L is the extent of the interaction region. From this expression we obtain the relationship between the force and the displacement of the atom:

$$F = -\alpha Mv_{\text{lon}}^2\rho/L = -\beta\rho \qquad (4.2)$$

Thus, we conclude that the force effecting the focusing of an atom in a beam should be proportional to the atomic displacement. This criterion must be satisfied for the force to produce the true "image": if this criteria is not met, the image is blurred, i.e. there are aberrations.

We could also consider the problem of the focusing of an atomic beam from point of view of the optical image theory: "*wave-optical approach*" [Papoulis, 1968]. In this approach the atomic beam should be considered in a form of de Broglie wave. In the optical image theory, the ideal objective lens is a transparency having the following phase transmission function:

$$T(x, y) = \exp\left[-ik(x^2 + y^2)/2f\right] \qquad (4.3)$$

where $k = 2\pi/\lambda$ and f is the focal length of the lens. A light beam passing through such a transparency undergoes an additional phase change of $k(x^2 + y^2)/2f$.

The task is to find such a potential field that would make the phase change of a wave function (de Broglie wave) to satisfy equation (4.3) in which k is equal now $2\pi/\lambda_B$ and λ_B is the de Broglie wavelength. It is known that if the de Broglie wavelength is small in comparison with the characteristic size conditioning a given problem, the characteristic of the system are close to classical. In the quasiclassical approximation, the atomic wave function is defined by the expression (Landay and Livshits, 1985):

$$\psi = \left(C/p(z)^{1/2}\right) \exp[(i/\hbar)\int p(z)dz] \qquad (4.4)$$

where C is a constant and $p(z) = [2M(E - U(z))]^{1/2}$ is the atomic momentum, M being the atomic mass, E the atomic total energy, and U(z) the atomic potential energy. In a quasi- resonant laser field ($\gamma \ll \delta = \omega - \omega_0$, $\delta \cong \gamma(I/I_s)^{1/2}$) the potential energy is given by (Gordon and Ashkin, 1980)

$$U = (\hbar\Omega/2)\ln(1 + s) \tag{4.5}$$

where s is an atomic transition saturation parameter.

The phase change of the wave function (4.4) due to the potential U(z) is

$$\delta\phi = \delta\phi_1 - \delta\phi_0 = (1/\hbar v_z) \int U(z)dz \tag{4.6}$$

where $\delta\phi_1$ and $\delta\phi_0$ are the phase changes with and without the laser field, respectively and v_z is the atomic velocity along the z-axis. By comparing the phase change (4.6) with the expression (4.3) for the transmission function of an ideal objective lens, it is possible to find the focal lens of the objective lens. To calculate the atomic beam density distribution in the focal plane, it can be used with the Kirchoff's diffraction theory. The effect of aberrations can be evaluated by considering the de Broglie wave front distortion caused by the objective lens.

Another *approach* to treat the focusing of atomic beam by laser light is to use the *path integral technique* to calculate the evolution of atomic wave function as it propagates in laser field [Gallatin and Gould, 1991]. In the path integral formalism, the solution of the Schrödinger equation is written in the form [Feynmann and Hibbs, 1965]:

$$\Psi(x_b, t_b) = \int d^3x_a K(x_b, t_b, x_a, t_a)\Psi(x_a, t_a) \tag{4.7}$$

where the propagator kernel $K(x_b, t_b, x_a, t_a)$ is given by a path integral:

$$K(x_b, t_b, x_a, t_a) = \int_a^b dx \, \exp\{[(i/\hbar)S(x(t))]\} \tag{4.8}$$

where the integration is over the all possible paths $x(t)$ that start at x_a at time t_a and end at x_b at time t_b.

The action $S(x)$ is evaluated along the path $x(t)$:

$$S(x) = \int_{t_a}^{t_b} dt\{(M/2)[dx/dt]^2 - U[x(t)] \tag{4.9}$$

where U is an effective potential energy of atom in laser field. In the quasiclassical approximation the kernel (4.8) has the form:

$$K \cong K_0 \, \exp[(i/\hbar)S_c] \tag{4.10}$$

where now S_c is evaluated along the classical trajectory of atom and K_0 is a normalization factor. The classical trajectory $x_0(t)$ can be found by the solution of equation of motion. The substitution of the classical trajectory in (4.10) the kernel (4.8) can be written in the form:

$$K(x_b, t_b, x_a, t_a) = \exp[(i/\hbar) \int_{t_a}^{t_b} dt \left\{ (M/2) \left[\mathbf{v}^2 - U \right] x_0(t) \right] \right\}} \tag{4.11}$$

Substituting (4.11) into (4.7), we obtain the approximate solution for the evolution of the atomic wave function Ψ.

The wave function for collimated atomic monochromatic atomic beam can be taken in the form:

$$\Psi(x_a, t_a) = \exp[-(i/\hbar)E t_a + ikZ_a]\phi(\mathbf{r}_a, z_a) \tag{4.12}$$

where $\hbar k$ is the initial atomic momentum, E_a is the initial energy and z is the axis of atomic beam. For the case of atomic beam be diffraction limited, the Gaussian profile for the function $\phi(\mathbf{r}_a, z_a)$ can be used for further calculations [Gallatin and Gould, 1991].

A *particle-optical approach* also can be used for treatment of focusing of atoms by laser light [McClelland and Scheinfein, 1991]. In this approach the atoms is treated as a classical particles that are moving in the potential field of a laser beam. This method was originally developed for charge particles optics for calculation of trajectories in cylindrically symmetric potential field. The equation of motion can be derived from Lagrangian, $L = v^2/2M - U(\rho, z)$, where $U(\rho, z)$ is the potential energy (4.5) and z is the axis of symmetry. In cylindrical coordinates, the radial equation of motion is:

$$\rho'' + (1/M)(dU(\rho, z)/d\rho) = 0 \tag{4.13}$$

By making assumption that potential energy less than kinetic energy of atoms $U(r, z) \ll E_0$, and $\rho' \ll 1$ (both conditions are usually valid in real experimental situation); the equation simplifies to:

$$d^2\rho/dz^2 + (1/2E_0)(dU/d\rho) = 0 \tag{4.14}$$

The quadratic radial dependence of the potential is a necessary condition for focusing of atoms. Expanding the real potential $U(\rho, z)$ and keeping only the lowest quadratic term in the expansion, we will find:

$$U(\rho, z) \cong k(z)\rho^2 \tag{4.15}$$

Equation (4.14) becomes:

$$d^2\rho/dz^2 + k(z)/2E_0\rho = 0 \qquad (4.16)$$

where $k(z)$ is determined by parameters of laser field.

The higher-order terms in the expansion (4.15) give a spherical aberration. The chromatical aberration can be found by calculations of trajectories of atoms with different initial kinetic energies. The diffusive spontaneous and dipole aberrations can be treated by adding in Eq. (4.16) a random force $F_r(t)$ for which $\langle F_r(t)\rangle = 0$ but $\langle F_r(t)F_r(t')\rangle \neq 0$. In [McClelland and Scheinfein, 1991] the explicit expressions for all aberrations for atom lens were obtained.

4.2 FOCUSING BY GRADIENT FORCE

The focusing of an atomic beam by means of the gradient force was demonstrated first at Bell Lab [Bjorkholm et al., 1978; Bjorkholm et al., 1980]). In their scheme, the atomic lens was created by CW dye laser which was focused to 200 μm and superimposed upon an atomic beam of sodium. The laser power was 50 mW and frequency detuning $\Omega = -2$ GHz. The atomic beam propagated along and inside a narrow near Gaussian laser beam. The laser frequency was tuned below the atomic transition frequency, so that the gradient force was directed toward the laser beam axis. The radial potential is determined by (4.5) with saturation parameter

$$s(z) = I/I_s\left[\gamma^2/\left(\gamma^2 + \Omega^2\right)\right]\exp\left[-2\rho/\rho_0^2(z)\right] \qquad (4.17)$$

In the experiment a width of atomic beam was compressed down to a spot diameter of 28 μm (Fig. 18). This minimum achievable spot diameter in the experiment was determined by a fluctuation of the momenta of atoms from spontaneous emission.

It is worth noting that in this experiment every atom scatters a small number of photons because of optical pumping so that the duration of resonance interaction atom with the field was less than the flight of the atoms through the laser beam. If the atoms interact with the field all the time, the transverse motion of the atoms will be periodical focusing and defocusing.

The problem of large thickness of previous atom lens (a large aberration) was overcome by group in Constanz University [Sleator et al., 1992]. In this work the atom lens was based on the large period standing wave (45 μm) produced by bouncing a laser beam off a glass surface under small angle. One period of such standing wave was used as a cylindrical lens for atoms (see Section 5). Atomic beam of metastable atoms of He* (25 μm diameter) crossed the standing waves at a perpendicular angle and was centered at the antinode of standing wave. The thickness of atomic lens was 80 μm

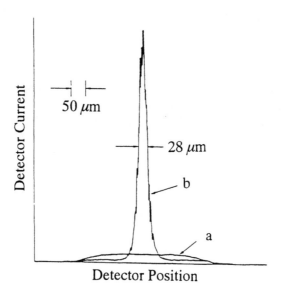

Fig. 18   Spatial profile of sodium atomic beam focused by gradient force [Bjorkholm et al., 1980].

resulting in an interaction time $t_{int} = 40$ ns is much shorter than the natural lifetime of excited state $2\,^3P_2$ of metastable He* ($\tau_{sp} = 100$ ns). The atomic beam was focused by such lens to a spot size of 4 μm. This lens was used also for imaging an atomic source which was formed by passing atoms through microfabricated transmission grating with a period 8 μm.

Further development of this idea of using a standing light wave was done in [Timp et al., 1992] where optical standing wave was used as an array of cylindrical lenses (each period of standing wave is a lens) to focus a perpendicular sodium atomic beam. The atomic beam was focused into the grating on the substrate with period of $\lambda/2$.

The most recent experimental work has been on applying the principle of laser focusing to deposition of chromium atoms on the surface. Figure 19(a) is a schematic of this experiment. A collimated, uniform atomic chromium beam is directed at a silicon surface. Grazing along the surface is an optical standing wave formed by retroreflecting a laser beam onto itself. Each nodes of standing wave acts as a cylindrical lens for atoms. The atoms are thus focused into a series lines with spacing equal to half of the wavelength. One-dimensional optical molasses in the atomic beam before it crosses the standing wave has been set up. In molasses region, the atoms have been cooled transversely to a temperature of 76 μK and an angular divergence of 0.3 mrad. This small angular divergence permitted a sharp focusing of the atoms. The chromium

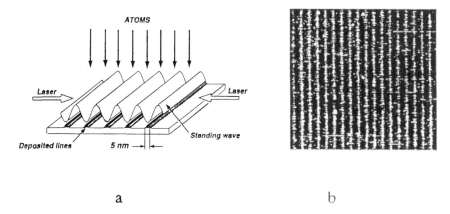

Fig. 19 a). Schematic representation focusing of chromium in standing wave. b). Image of Cr lines on Si substate. Lines are 50 nm wide, spread by 212 nm.

atoms were deposited on the silicon substrate and observed with either a scanning electron microscope or an atomic force microscope. Figure 19(b) shows the image of Cr lines. The widths of the lines are about 50 nm.

4.3 FOCUSING WITH ATOM LENS BASED ON LIGHT PRESSURE FORCE

Another light field configuration for the focusing of an atomic beam is based on using a light pressure force and it was demonstrated in the Institute of Spectroscopy in Moscow [Balykin et al., 1986; Balykin et al., 1988a]. The atomic lens was formed by divergent Gaussian beams propagating pairwise in opposite direction along the x- and y-axis perpendicular to the atomic beam, Fig. 20. The waist of these beams are situated at equal distances from the center of atomic beam. The lasers were tuned to precise resonance with the atomic absorption frequency. Under such a condition an atom, moving away from the atomic beam axis, is acted upon a light pressure force (Fig. 17) that tends to bring it back to the beam axis. The effect of gradient force in this case is insignificant. As we will see below, such a configuration could be an atom lens for a beam of neutral atoms.

4.3.1 One dimensional laser lens: focusing and imaging

The simplest lens based on light pressure force is one dimensional lens formed by two elliptical Gaussian laser beams propagating in opposite direction along the x-axis.

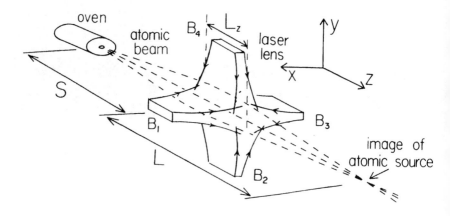

Fig. 20   Four divergent Gaussian laser beam used to form atom lens.

Such lens is similar to a cylindrical optical lens. Let us consider the basic property of such a lens. The light pressure force acting on atoms may be expressed as [Balykin et al., 1988a]

$$\mathbf{F} = 8\hbar k\gamma \left[G_0 l / \left(1 + 2G_0 + \Omega^2/\gamma^2 \right) \right]$$

$$\left\{ \left[-\left(\rho_{oz}^2/\rho_z^2 b_z^2 \right) + \left(\rho_{o^2}/\rho^2 b^2 \right) \right] x e_x + \left(\rho_0^2/b^2 \right) y e_y \right\} \qquad (4.18)$$

where ρ_0 and ρ_{oz} are the radii of the elliptical Gaussian beams at the waists of a beam, ρ and ρ_z are the radii of the same beam at the origin of coordinates, $b = 2k\rho_0^2$; $b_z = 2k\rho_{oz}^2$, l — is the distance from the waist to the origin of coordinates. It can be seen from (4.18) that on the x-axis the force is proportional to the atomic displacement and is directed towards the z-axis, while on the y-axis the force is an expulsive force which will broaden the atomic beam in the direction of y-axis, but nevertheless will not prevent the image on the x-axis being obtained. From the expression (4.18) we can see also that the force on the x-axis depends on the laser divergence along the z-axis and the y-axis.

Let us now determine the basic atom lens parameters. We proceed from the laws governing the motion of an atom under the action of the radiation force. The relationships obtained are considerably simplified if the following two approximations are used: (i) an atom interacts with the radiation only in a strictly defined region of length z_0, in which the force has the form (4.18) and (ii) the atomic displacement from the z axis in the interaction region varies only slightly, so that the force is constant and is determined only by initial displacement ρ_0. These approximations do not essentially alter the basic results but allow the equation of motion to be easily integrated and analytical expressions obtained for the displacement ρ and transverse velocity of atom v_ρ:

$$\rho = \rho_0 + v_{\rho 0}\tau_{int} + F\tau_{int}^2/M \tag{4.19}$$

$$v_\rho = v_{\rho 0} + F\tau_{int}/M \tag{4.20}$$

where $\tau_{int} = z_0/v_{long}$ is the time of atomic-radiation interaction, ρ_0 and $v_{\rho 0}$ are the initial atomic displacement and transverse velocity at the entrance to the lens, M is the atomic mass and v_{long} is the longitudinal velocity.

Using the equations (4.19) and (4.20), one can derive the following expression for the focal length of the laser lens:

$$f = \left(v_{long}^2/f_o^2 z_0\right)\left[\left(1 + 4G_0 + \Omega^2/\gamma^2\right)/G_0\right] \tag{4.21}$$

where

$$f_o = \left[(8\hbar k\gamma/M)\left(1q_{oz}^2/q_z^2 b_z^2\right)\right]^{1/2} \tag{4.22}$$

With the laser radiation and atomic beam of Na having the parameters $G_0 = 0.3$, $\Omega = 0, z_0 = 3\,cm, l = 1\,mm, k = 10^5$, the focal lenth of one-dimensional lens is estimated to be 3 cm for atoms with the velocity $v_{long} = 5 * 10^4$ cm/s. An atom lens with such a focal length can easily be investigated using an atomic beam with thermal velocity.

The atom lens, like its counterpart in geometrical optics, suffers from aberrations. It is evident from (4.21) that the focal length is proportional to the square of the longitudinal atomic velocity. This leads to the chromatic aberrations. Since the atomic beam has the Maxwell velocity distribution, atoms with different atomic velocities will be focused at different points.

Another source of aberration is the finite cross-sectional dimension of the atomic beam. A violation of the requirement that the atomic trajectories are paraxial ($\rho \ll 1$) results in additional terms that are not proportional to the displacement ρ from the z axis which appear in the expression for the radiation force (4.18). This causes spherical aberrations.

The question arises which resolution one can expect to achieve with atom lens based on the light pressure force. In light optics the resolution power of a lens is determined by diffraction of light at the lens aperture edges. The maximum possible resolution of atom lens based on the light pressure force is governed mainly by momentum diffusion of atoms in the course of their spontaneous reemission of photons. The minimum spot radius of an atomic beam in the image plane restricted by momentum diffusion is

$$\delta\rho = \langle\delta v_{dif}\rangle/\tau_{int} \qquad (4.23)$$

where $\langle\delta v_{dif}\rangle$ is the deviation of the transverse atomic velocity by the diffusion process and τ_{int} is the time of flight of the atom through the laser lens. The spread of transverse velocity $\langle\delta v_{dif}\rangle$ may be expressed in terms of momentum diffusion coefficient D as follows:

$$\langle\delta v_{dif}\rangle = \left(2D\tau_{int}/M^2\right)^{1/2} \qquad (4.24)$$

At resonance and with the saturation parameter G = 1, the momentum diffusion coefficient is

$$2D \cong \hbar^2 k^2 \gamma \qquad (4.25)$$

Using (4.23)–(4.25), we find that the momentum diffusion-limited "image" spot radius in the atom lens is

$$\delta\rho_{min} = (2\hbar^2 k^2 \gamma q_z/M^2 v_l)^{1/2} \, (f/v_l) \qquad (4.26)$$

For the laser and atomic beam parameters considered above, the minimum "image" sport radius is $\rho_{min} = 60 \ \mu m$.

Let us briefly consider the experiment [Balykin et al., 1986; Balykin et al., 1988a] in which was achieved the focusing of atomic beam by light pressure force. In the experiment was observed also the imaging of a source of atomic beam. The atomic lens was formed by two divergent Gaussian laser beams whose waist were at a distance 2 mm from the symmetry axis of the lens. The atom-radiation interaction length was 10 mm. The beam of sodium atoms was formed by one exit hole for focusing of the atomic beam or two exit holes separated by 2 mm symmetrically about the atomic beam axis for the imaging of the atomic source. To observe of the focusing or the imaging, the spatial distribution of the atomic density in the beam was registered by observing the fluorescent signal from single frequency probe laser beam as a function of the transverse atomic coordinate. The diameter of probe laser beam was considerably less than the size of atomic beam in image plate. This laser beam moved parallel itself and crossed the atomic beam at certain region along the axis of the atomic beam. The probe laser frequency was tuned to resonance with D_2 transitions in a certain portion of the Doppler counter of the atomic beam so as to intercept atoms with a definite velocity.

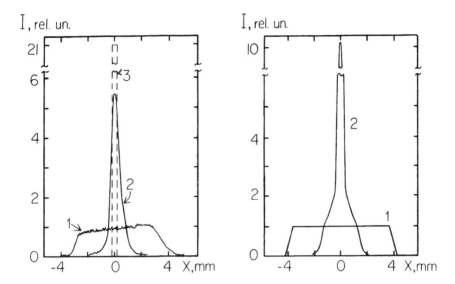

Fig. 21 Focusing of atomic beam by atom lens based on the light pressure force. Profiles of atomic beam in "image" plate. Curve 1 shows the measured atomic intensity in a beam in the absence of laser field. Curve 3 presents the beam profiles in the case of its ideal focusing, calculated by the law of geometrical optics. Curve 2 shows the measured profile of the focused atomic beam. On the right side — the calculated profile of atomic beam (1 — initial profile, 2 — after focusing) [Balykin et al., 1986].

Figure 21 shows the focusing of the beam of sodium atoms by the atomic lens. In the experiment the longitudinal atomic velocity was $3.8\ 10^4$ cm/s. Curve 1 shows the measured atomic intensity in a beam in the absence of laser field. Curve 3 presents the beam profiles in the case of its ideal focusing calculated by the law of geometrical optics. Curve 2 shows the measured profile of the focused atomic beam. The atomic intensity in the beam center as the result of focusing was increased by a factor 5.5. The distortion of the "image" was mainly due to the spherical aberration.

The imaging by means of atom lens of a source having more complex configuration, for example, one with two holes, is a more interesting task. Such an atomic source would correspond to a two point source in a light optics. Figure 22 demonstrates the image of two atomic beam point sources by a atomic lens. The density distribution of atomic "image" consists of two peaks which corresponds to the two holes in atomic source.

We have already noted that the atomic lens suffers from chromatic aberration. In this experiment the effect of chromatic aberration was measured on the formation

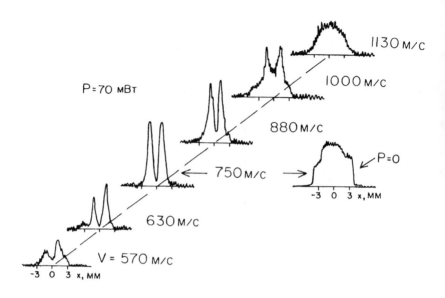

Fig. 22   Image of two atomic source. Atomic beam profiles in the image plate as function of the longitudinal atomic velocity. On the right side — the profile in the image plate without light. (Balykin et al., 1988a).

of the source "image" and on the resolution of the atomic lens. For this purpose the beam profile was registered after focusing at a definite, fixed distance from the laser lens and the longitudinal velocity of atoms being registered was varied by altering the position of probe laser frequency within the Doppler absorption counter of the atomic beam (Fig. 22). The original profile of the beam of atoms with a velocity $7.5 * 10^4$ cm/s is shown on the right side of the picture. Shown on the left side of the figure are the focused atomic beam profiles, arranged in a line, from the two sources. The longitudinal velocity was varied from $5.7 * 10^4$ cm/s to $11.3 * 10^4$ cm/s. One can appreciate qualitatively that with fixed laser power, there should be a group of atoms with optimum longitudinal velocity, which are focused precisely into the registered region, the slower atoms being focused before and the faster ones, after this region. That is to say, there should be some optimum velocity at which the "image" of the two sources will be defined most sharply. This can be seen exactly in Fig. 22. At velocities below $5.7 * 10^4$ cm/s no satisfactory focusing is observed at all. As the longitudinal atomic velocity was increased, there appears a two peak structure corresponding to the focusing of the atomic beam from the two sources. The best resolution was observed at velocity equal to $7.5 * 10^4$ cm/s. As the atomic velocity further increased, the resolution of the laser lens decreases until, finally, the two peak structure vanishes.

4.3.2 Two-dimensional atom lens

The two-dimensional atom lens is shown schematically in Fig. 20. It is formed by four divergent Gaussian beams propagating pairwise in opposite directions along the x- and y-axis. In this laser lens the light pressure force exerted on atom can be represented as [Balykin et al., 1989b]:

$$\mathbf{F} = \mathbf{F}_\rho + \mathbf{F}_z \qquad (4.27)$$

where $\mathbf{F}_\rho$ is the radial part of force and $\mathbf{F}_z$ is the axial force. The radial force $\mathbf{F}_\rho$ is responsible for the focusing of atomic beam; the axial force $\mathbf{F}_z$ is along the atomic beam axis. The focusing ot the beam can be achieved if radial force $\mathbf{F}_\rho$ is larger than the $\mathbf{F}_z$ force. The radial force increases on the increase in the divergence of the laser beam, because the divergence determines the unbalance of the intensities in the optical field. A considerable divergence can ensure that the radial part of force is larger than the axial force. The dependence of the radial force on the z coordinate shows that the force is a variable sign quantity: along the edges of the laser lens it pulls the atoms inside (focusing), whereas at the center it pushes them out (defocusing). The overall effect of the force on the atom at the edges and the center can be focusing or defocusing, depending on the divergence of laser beam [Balykin et al., 1989b] and the positions of atomic source and image plate.

4.4 THE POSSIBILITY OF DEEP LASER FOCUSING OF ATOMIC BEAM

A serious drawback of previous discussed atomic lens is its insufficient resolution. The minimum spot size in Bell Labs focusing experiment is about 30 μm. The minimum resolution of an atomic lens based on the radiation pressure force is also of the same order of magnitude. This resolution limit is due to momentum diffusion, which smears the trajectories of the atoms interacting with the laser lens field. In the Constanz experiment the minimum sport size is considerably smaller but the focusing is only one-dimensional one.

Resolution can be considerably improved by using the same idea of the gradient force but now with a different laser field configuration and a different atom-field interaction geometry as was proposed in [Balykin and Letokhov, 1987b]. A new atomic objective lens is a focused TEM^*_{01} laser beam tuned above resonance, Fig. 23. The atomic beam propagates along the lens axis, where the intensity of light and therefore the rate of spontaneous emission and momentum diffusion, is minimum. This solves the diffusion problem. This configuration was analyzed in the thin lens approximation in [Balykin and Letokhov, 1987b]. Later the path-integral techniques

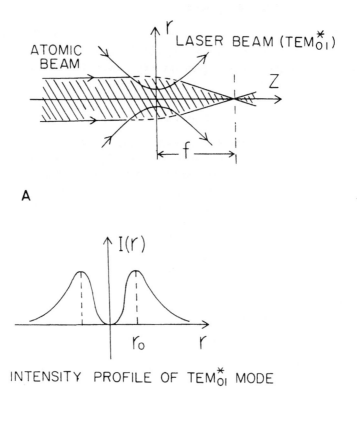

Fig. 23 Laser field configuration for deep focusing of atomic beam. (A) disposition of laser and atomic beam; (B) cross-sectional intensity profile in the TEM_{01}^* laser mode field (Balykin and Letokhov, 1987b).

[Gallatin and Gould, 1991] and methods developed for particle optics [McClelland and Scheinfeind, 1991] were used to achieve a more general result. Let us consider the basic properties of such objective lens.

To realize focusing, it is necessary to have a laser beam providing for a change of the wave function a phase shift equal to the argument of the exponential in eq. (4.3). This purpose can be served by the TEM_{01}^* laser mode near the beam axis. The laser intensity in the TEM* mode is defined as [Rigrod, 1963] (Fig. 23).

$$I(x, y) = 4I_0 \left[(\rho_0^0)^2/\rho_0^2(z) \right] \left[2\rho^2/\rho_0^2(z) \right] \exp\left[\left(-2\rho^2/\rho_0^2(z) \right) \right] \qquad (4.28)$$

where ρ_0^0 is the laser beam waist radius in the plane $z = 0$, $\rho_0(z)$ is the beam waist radius in the z plane, and $I_0 = P_0/2\pi(\rho_0^0)^2$, P_0 — being the radiation power. In the case of paraxial optics ($\zeta = \rho/\rho_0 \ll 1$) the expression for the phase change of a de Broglie wave in a laser field of the form (4.28) has the following form:

$$\delta\phi = \left(\pi^2 \Omega z_R \alpha/v_z \right) \zeta^2 \left[1 - (1 - \alpha/2)\zeta^2/2 + (3/16)(1 - 4\alpha)\zeta^4 \right] \qquad (4.29)$$

where $\alpha = \left(P_0 \gamma^2 \right) \left(\pi I_s \Omega^2 (\rho_0^0)^2 \right)$.

The phase change turns out to be proportional to the square of the radius ζ^2. To the first approximation, the bracketed terms proportional to ζ^2 and ζ^4 can be neglected. In such approximation the field (4.28) acts on atomic beam as an ideal objective lens with the resolution determined by its numerical aperture and the de Broglie wavelength. The bracket terms describe spherical aberration of the 4th and the 6th order, respectively.

By comparing the phase change (4.29) with the expression (4.3) for the transmission function of an ideal objective lens, it is easy to find the focal lens of the objective lens

$$f = \hbar (\lambda\Omega/P_0) \left(\rho_0^2 \right) \left(I_s/\gamma^2 M\lambda_B^2 \right) \qquad (4.30)$$

The expression in the first parentheses is determined by the laser beam parameters, that in the second by the beam waist, and in the third by the atomic parameters.

The resolution of a atomic objective lens depends mainly on the following factors: a) a momentum diffusion caused by spontaneously re-emitted photons, b) a chromatic aberration due to the strong atomic velocity dependence of the focal length, and c) a spherical aberration inevitable where a real laser field is used, d) a diffraction of atoms by the aperture limiting the cross-sectional size of atomic beam, e) a fluctuation in dipole force. Let us consider how strongly these factors affect the resolution of the objective lens.

Diffusion aberration. An atom interacting with a quasi-resonance laser field can absorb and then spontaneously reemit photons. Assuming a spherical symmetry of the reemmited photons, the change of transverse atomic momentum is

$$\delta p = (2N/3)^{1/2}\hbar k = \hbar\delta k_B^\perp \qquad (4.31)$$

where $\delta k_B^\perp$ is the change of the transverse de Broglie wave vector component. If the atomic beam diameter is equal to $2\sigma_0$ we find that the transverse interval of the wave vector variation due to diffraction is

$$\delta k_B^0 \cong 1/2\sigma_0 \tag{4.32}$$

The diffusion aberration can be neglected if

$$\delta k_B^\perp < \delta k_B^0 \tag{4.33}$$

The relation (4.33) is equivalent to the limitation of the number of spontaneously reemitted photons:

$$N < (\lambda/2\pi\sigma_0)^2 \tag{4.34}$$

From condition (4.34) we can get the following condition under which the diffusion aberration can be neglected [Balykin and Letokhov, 1988d]:

$$\Omega^2 > \left(8\pi^4 P_0 \gamma^3 \rho_0^2 / 3 I_s v_z \lambda^3\right) \left(\sigma_0^4 / \rho_0^4\right) \tag{4.35}$$

Chromatic aberration. The effect of chromatic aberrations can be evaluated by considering the de Broglie wave front distortion caused by it. It the Rayleigh quarter-wave criterion is used as an image-quality criterion, the following restriction is imposed upon the nonchromatic character of atomic beam:

$$\delta v/v < \left(32^{1/2}/6\pi\right) \left(f/\sigma^2\right) \lambda \tag{4.36}$$

The contribution to the spot diameter in the image plane from this type of aberration is [Gallatin and Gould, 1991]:

$$\sigma_{ch} \cong 2\sigma_0 (\delta v/v) \tag{4.37}$$

where σ_0 is the atomic beam diameter (the entrance aperture of atomic beam into the lens).

Spherical aberration. It can be seen from the expression for the phase shift (4.29) that the effect of spherical aberration can be eliminated by reducing the atomic beam diameter. But this way leads to an increased role of diffraction and reduced beam intensity at the center of the diffraction pattern. The spherical aberration can be evaluated by calculating the change in the spot diameter of atomic beam in image plane. In [Balykin and Letokhov, 1988d] the sport diameter was calculated by using Kirchoff's diffraction theory. [Gallatin and Gould, 1991] evaluated the spherical aberration by calculating the change in the sport diameter owing to the shift in focal position for off-axis trajectories of atoms by using the path-integral techniques. The calculated spot radius is:

$$\sigma_S \cong (\sigma_0/\rho_0)^2 \sigma_0 \qquad (4.37)$$

For $\sigma_0/\rho_0 = 0.04$ the atomic spot radius in image plate may be as small as about 10 Å [Balykin and Letokhov, 1988d; Gallatin and Gould, 1991].

Diffraction. Since the potential in atomic lens is slowly varying on the scale of de Broglie wavelength, we may apply the quasi-classical approximation and make analogy of diffraction treatment of atomic beam with the diffraction treatment in ordinary optics. For atomic beam with a constant spatial density distribution and circular aperture of radius σ_0, limiting the size of atomic lens the expression for beam size at image plate is

$$\sigma_{\text{diff}} \cong 0.6\, \lambda_{\text{Br}}(f/\sigma_0) \qquad (4.39)$$

where f is the focal length of the lens.

Fluctuation in the dipole force. Another source of diffusion aberration is the fluctuation in the dipole force (Dalibard and Cohen-Tannoudji, 1985). The contribution to the sport diameter from this type of diffusion is [Gallatin and Gould, 1991]:

$$\sigma_d \cong (v\lambda/\gamma)^{1/2} (\sigma_0/\rho_0)^3 \qquad (4.40)$$

The contribution trom this type of aberration is usually less than that from another ones.

To calculate the atomic density distribution in the focal plane, the Kirchoff's diffraction theory was used in [Balykin and Letokhov, 1988d]. Figure 24 illustrates the atomic beam distribution in the focal plane of the objective atom lens formed by laser beam focused in spot $\rho_0 = \lambda$. The solid curve corresponds to aberration-free case (the diffracted limited spot diameter), dashed curve shows the distribution with the chromatic aberration, dashed-and-dotted curved represents that with spherical aberration allowed for. The distribution calculated with a diffusion aberration coincides with that in the aberration free case. The curves were calculated for the following parameters the laser and atomic beams: the laser power $P = 1W$, the velocity of atoms $v = 2.2 * 10^5$ cm/s, the ratio of the atomic beam to the waist of laser beam is 0.25 and $\delta v/v = 10^{-3}$. It can be seen from the figure that with aberration taken into account the atomic beam size at the focal point differs not very greatly from the diffracted limited spot diameter.

In conclusion of this paragraph, let us list the requirement that must be met by the laser radiation and by the atomic beam to be able a deep focusing of an atomic beam. The focusing potential field is produced by using the TEM_{01}^* laser

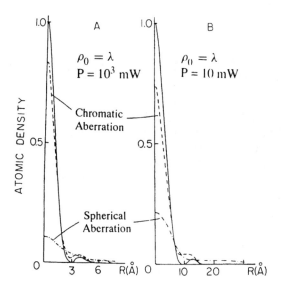

Fig. 24 Atomic beam distribution in the focal plane of atom lens based on the TEM^*_{01} laser mode field. The solid curve corresponds to aberration-free case, dashed curve shows the distribution with the chromatic aberration, dashed- and dotted-curves represent that with spherical aberration allowed for. The distribution calculated with diffusion aberration coincides with that in the aberration free case. The minimum waist of laser beam $\rho_0 = \lambda$ [Balykin and Letokhov, 1987b].

mode strongly focused to a size of the order of wavelength of light. The radiation power needed to focus beams having thermal velocities is several hundred milliwatts. Diffraction resolution of the atomic objective could be realized at an atomic-beam monochromaticy $\delta v/v = 10^{-3}$.

Using the atomic beam deep focusing technique considered here, it is not difficult to conceive an atomic beam microscope similar to a reflected or transmission scanning electron microscope.

4.5 OTHER TECHNIQUES

Focusing of atomic beam can also be reached by using other methods (see Introduction). We will briefly mention the latest achievement by these method. One of the first scheme of focusing of atomic beam was based on the interaction of a magnetic or electrical dipole moments of atoms and molecules with a static field: hexapole

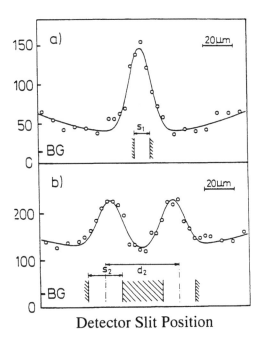

Detector Slit Position

Fig. 25 Focusing of atom by Fresnel zone plate. Atomic density distribution in the image plane of metastable helium atoms: (a) image of single slit and (b) double slit (Carnal et al., 1991).

magnets [Friedburg et al., 1950, 1951; Friedburg and Paul, 1950, 1951; Korsynskii and Fogel, 1951; Vanthier, 1949], (quadrupole electrostatic lenses [Benewitz, 1955; Gordon et al., 1955].

One of the latest achievements in the focusing of atom by their interaction with matter is the reflection of hydrogen beams at a liquid-helium-vacuum interface [Berkhout et al., 1989]. The mirror consists of fused-quartz substrate polished to optical precision and coated with liquid ^{4}He film to obtain high reflectivity. Reflection of He atom at liquid-vacuum interface has been observed earlier [see Nayak et al., 1983] however, the high reflection was not achieved except at grazing incidence. Berkhout et al. [1989] demonstrated focusing of a high divergent atomic beam at normal incidence. The measured reflectivity was of 80% and determined by a static surface roughness due to substrate and dynamic surface roughness due to helium film.

Although zone plates have been already recognized as a focusing element in 19th century, they have found very little application in classical optics. In recent years microfabricated Fresnel zone plates have been applied in X-ray microscopy [Schmahl and Rudolph, 1984] and in the focusing of slow neutron beam [Kearney et al., 1980]. Recently, free standing microfabricated structures have successfully been used for the diffraction of atoms [Keith et al., 1988; Carnal et al., 1991] and also applied in the first atomic interferometers [Carnal and Mlynek, 1991; Keith et al., 1991]. In Constanz University the Fresnel zone plate was at the first time used also as a focusing device for atoms [Carnal et al., 1991]. In a focusing experiment atoms of metastable helium with atomic de Broglie wavelength $\lambda_{Br} = 0.5\text{–}2.5$ Å are passed through either a single or a double slit with dimension in the $10 \, \mu$m range. The Fresnel zone plate was $210 \, \mu$m in diameter and with an innermost zone diameter of $18 \, \mu$m. For $\lambda_{dB} = 1.96$ Å the focal lens of Fresnel zone plate is 0.45 m. Figure 25 shows the image (a) of single slit and (b) double slit. The observed peaks corresponding to the slits can be clearly identified in the picture.

5. CHANNELING OF ATOMS IN A STANDING WAVE

There are several reasons to the interest of interaction of atoms with a standing wave. First, this is another example of atomic interaction with a particular simple laser field configuration where nowadays is a satisfactory agreement between a theory and an experiment. Second, the interaction of atoms with a standing wave give rise to such interesting effects as a diffraction and a channeling of atoms. The diffraction of atoms by standing wave was considered in the several reviews [Kazantsev et al., 1985; Stenholm, 1986] and the special issues [Meystre and Stenholm, eds., 1985; Chu and Wieman, eds., 1991; Mlynek, Balykin and Meystre, eds., 1992].

The idea of atomic channeling in a standing wave was set first in [Letokhov, 1968], the authors of which suggested that the effect should be used to eliminate the Doppler broadening of spectral lines by Dicke narrowing [Dicke, 1953]. The effect was indeed observed by the group National Institute of Standards and Technology, Gaitherburg in the fluorescent spectrum of Na in 3D optical molasses [Westbrook et al., 1990] and later even the transition between vibrational sublevel of the trapped atoms in 1D optical molasses were measured [Jessen et al., 1992; Verkerk et al., 1992].

Here we are going to give a brief review of the recent theoretical and experimental results of channeling of atoms in the standing wave. We consider the channeling (one dimensional localization) of atoms in two configurations of the laser fields: a monochromatic and a bichromatic one dimensional laser fields. A quasi resonant monochromatic laser field creates a spatially periodic potential field with a period equal to a half of the wavelength. If the atomic kinetic energy is less than the height of potential wells the atom can be trapped in such a field. In a bichromatic standing wave the spatial size of the potential wells could be larger than the wavelength due to effect of "rectification" of the gradient force [Kazantsev and Krasnov, 1987] and localization can be made on a macroscopic scale.

5.1 SINGLE FREQUENCY LASER FIELD. THEORETICAL CONSIDERATION

5.1.1 Atomic potential in standing wave. "Dressed" state approach

Let us consider a two level atom with a ground state $< g|$ and excited state $< e|$ in laser field with a frequency ω_L. The uncoupled states of atom + photons of laser field can be written as $|g, n + 1 >$ and $|e, n >$. The first state corresponds to atom in the initial state $|g >$ in the presence of $n + 1$ photons, the second state corresponds to atom in the excited state $|e >$ and the presence of n photons. These states are bunched in manifold separated by energy $h\Omega$. When the coupling is taken into account, the

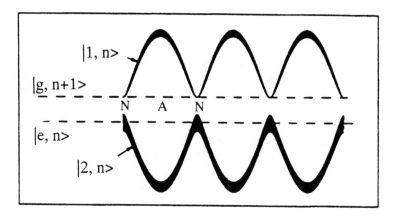

Fig. 26 The spatial variation of the energy of dressed states in one-dimensional standing wave. At the nodes of standing wave (where an electrical field is zero and as a consequence $\omega_R = 0$) the two dressed states coincide with unperturbed ones. At an antinodes the shift of energy takes its maximum value [Dalibard et al., 1987].

two unperturbed states transformed into perturbed states $|1, n>$, and $|2, n+1>$. The bunches of these states are called "*dressed*" states [Dalibard and Cohen-Tannoudji, 1985]. Each dressed state is a linear superposition of the unperturbed state $|g, n+1>$ and $|e,n>$. The energies of atom in the dressed states are:

$$E_1 = (n+1)\hbar\omega_L - \hbar\Omega/2 + \hbar\Omega_R/2 \qquad (5.1)$$

$$E_2 = (n+1)\hbar\omega_L - \hbar\Omega/2 - \hbar\Omega_R/2 \qquad (5.2)$$

where Ω_R is off-resonance Rabi frequency. The dressed state is now separated by $\hbar\Omega_R = \hbar(\Omega^2 + \omega_R^2)^{1/2}$. The last term in expression (5.1) and (5.2) is related to the interaction of atom with a laser field and can be interpreted as a potential energy of atom in a laser field. In inhomogeneous laser beam, these energy are varied with a position of atom: $\omega_R = \omega_R(r)$. Figure 26 shows the spatial variation of the energy of dressed states in one-dimensional standing wave. At the nodes of standing wave (where an electrical field is zero and as a consequence $\omega_R = 0$) the two dressed states coincide with unperturbed ones. At an antinodes the shift of energy takes its maximum value.

If we take into account the coupling of the dressed states with a vacuum field, then through spontaneous emission both states $|1, n>$ and $|2, n>$ can decay to the state $|1, n-1>$ and $|2, n-1>$: i.e. the atom changes its energy from E_1 to E_2.

Spontaneous emission transfers "atom + laser field system" from one dressed state to another one. In this case atom is moving in a "mean" potential:

$$U = (\hbar\Omega_R/2)\,n_1 - (\hbar\Omega_R/2n_2) \tag{5.3}$$

where n_1 and n_2 are the dressed state populations This expression for potential energy can be written in familiar form (see (2.23)):

$$U = (\hbar\Omega/2)\ln(1 + s) \tag{5.4}$$

where s is the off-resonance saturation parameter (2.23b). From (5.3) it is clear that the "mean" potential is less than a potential of atom in a certain dressed state. The force determined by this potential is called a gradient force (see Section 2).

A spontaneous emission causes not only coupling dressed states and decreasing the potential energy of atom but also give rise to a velocity dependence of gradient force. The velocity dependent part of a gradient force is called a stimulated (retarded) force and it was considered in Section 2.

5.1.2 Lifetime of atom in a single potential well

The motion of atoms in a standing wave is governed by the gradient force, the friction force and momentum diffusion. When the atomic transition saturation parameter is much less than unity (s < 1) and the field-atom interaction time is such that the change in atomic momentum due to the friction and the diffusion is insignificant, both the friction force and momentum diffusion can be disregarded. In this case, an atom sees the light wave as a spatially periodic potential field, the period of which is equal to that of the spatial field intensity distribution, i.e.

$$U(z) = -U_0\cos 2kz, \quad U_0 = \hbar\omega_R^2/2\Omega \tag{5.5}$$

An atomic ensemble placed in a potential field (5.5) is divided in two groups. The first group consists of the atoms with a total energy of $W = E(z) + U(z) < U_0$ ($E(z)$ is the kinetic energy of the atoms). These atoms reside between two adjacent maxima of the potential $U(z)$, the distance between which is $\lambda/2$ i.e., they are spatially localized. The localized atoms perform a harmonic oscillations with an amplitude less than $\lambda/2$ about the potential minima. The second group includes those atoms for which $W > U_0$. These atoms have an infinite motion, i.e., they are not localized. If the entire atomic ensemble has a continuous energy distribution, the localized and nonlocalized atoms are spatially intermixed.

In the case of weak saturation of the atomic transition ($G < (1 + \Omega^2/\gamma^2)$) and in the limit of large detuning ($\Omega \gg kv$) the total force has, to the first order in kv/Ω, the form [Minogin and Rozhdestvensky, 1987]:

$$F = F_{fr} + F_{osc} = 4\hbar k^2 \left(\omega_R^2 \gamma / \Omega^3\right) v_z \sin^2 kz + \hbar k \left(\omega_R^2 / \Omega\right) \sin 2kz \quad (5.6)$$

The second part of this force is the gradient force which corresponds to the potential (5.4), the first part is the friction force. If a cold atom is placed in one minimum of the potential, it will be localized in this potential well until its kinetic energy attains, as a result of diffusion heating, value equal to the depth U_0 of the potential well. The friction force does not stabilize the atom in the vicinity of the bottom of the well, since F_{fr} is small in the vicinity of the bottom and equal to zero at the bottom.

Using the diffusional law of increase of the kinetic energy of the atom, the time of localization of atom in the potential U can be expressed as:

$$t_{loc} = \langle(\delta p)^2\rangle / D \quad (5.7)$$

where $\langle(\delta p)^2\rangle = 2MU_0$. The coefficient of the diffusion at the minimum is given by the formula [Minogin and Rozhdestvensky, 1987]:

$$D \approx \hbar^2 k^2 \gamma \omega_R^2 / 2\Omega^2 \quad (5.8)$$

From (5.7) and (5.8) we have:

$$t_{loc} \approx (\hbar/2R)(\Omega/\gamma) \quad (5.9)$$

where $R = \hbar^2 k^2 / 2M$ is the recoil energy. From (5.9) we see that the time of localization of atoms in a standing wave is always increased when the detuning is increased, but the depth of the well U_0 also is decreased. The maximum depth of the potential well is attained at $\omega_R \approx \Omega$. In this case the localization time is of the order of

$$t_{loc} \approx (\hbar/2R)(G/2)^{1/2} \quad (5.10)$$

For example, for the case of the saturation parameter $G = 10^8$ and $\omega_R \approx \Omega = 10^4\gamma$, the localization time of atom is $t_{loc} \sim 1$ sec.

5.1.3 Atomic trajectories in one dimensional standing wave

Let us now consider the motion of atoms in the simplest standing wave: one-dimensional monochromatic standing wave. In this case atoms can be localized in one direction and will move freely in another direction: a *channeling* of atoms along the wavefront of light wave. The atoms entering the standing wave is divided in two groups. The first group contains the atoms with a total energy less than the kinetic energy of the atoms; they are spatially localized. The second group includes those atoms for which the kinetic energy is larger than the potential energy. They are not localized. In the considered case, the motion of atoms can be calculated by using a quasi-classical approximation. In this approximation the problem is reduced to the integration of Langevin equation [Balykin et al., 1989c]:

$$dp/dt = F \tag{5.11}$$

where p is the atomic momentum and F is the total force acting on the atom, which may be represented in the form

$$F = F_g + F_{fr} + F_d \tag{5.12}$$

where F_g is the gradient force, F_{fr} is that part of the force that depends on the atomic velocity (i.e. the friction force), and F_d is the fluctuating force that is due to the atomic momentum diffusion.

The gradient force for two level atoms is given by [Gordon and Ashkin, 1980]

$$F_g = \hbar\Omega[s/(1+s)]\alpha \tag{5.13}$$

where

$$s = \omega_R^2/2\left(\gamma^2 + \Omega^2\right) \tag{5.14}$$

is the local off-resonance saturation parameter, and

$$\alpha = \text{grad ln } u \tag{5.15}$$

and u is the local field amplitude in the standing wave.

The friction force for two-level atom is expressed as

$$F_f = \hbar\Omega\left[s/(1+s)^3\right]\left\{\left[2\gamma^2(1-s) - s^2\left(\Omega^2 + \gamma^2\right)\right]\Big/\gamma\left(\gamma^2 + \Omega^2\right)\right\}^*$$

$$\left\{1/\left[1 + A(\Omega, s)(\alpha v/\gamma)^2\right]\right\}(v\alpha)\alpha = -\eta(vn)n \tag{5.16}$$

where $n = \alpha//\alpha/$ and η is the friction coefficient.

The last term in equation (5.12) is the force due to momentum diffusion in both magnitude and direction, the fluctuation of momentum during the time δt being given by

$$\delta p = (2D_1\delta t)^{1/2}\xi_1 n + (2D_2\delta t)^{1/2}\xi_2 u \tag{5.17}$$

66 V.I. BALYKIN AND V.S. LETOKHOV

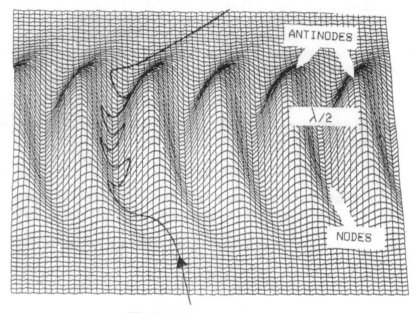

ANTINODES

λ/2

NODES

TRAJECTORY OF ATOM

Fig. 27 Localization (channeling) of atoms in a one-dimensional spherical standing wave. The picture shows the atomic potential energy in spherical standing wave. The curves reflect the trajectories of localized atom [Balykin et al., 1989c].

where ξ_1 and ξ_2 are the Gaussian random numbers, $\mathbf{u}$ is a random unit vector, D_1 is the directional diffusion coefficient, and D_2 is the anisotropic diffusion coefficient (see, Section 2).

Let us now consider a two-dimensional atomic motion in a standing wave formed by a laser beam with the Gaussian intensity distribution, Fig. 27. Suppose that the atomic beam transverses the standing wave. At a certain point of the standing wave where the transverse kinetic energy equals the potential energy of the atom in the field, atoms can be localized in potential well. After that atoms will be channeled along the node (or loop) of the light wave: the atomic trajectory must follow the wavefront accurately to within $< \lambda$ (Figure 27). Figure 28, shows the behavior of the potential energy of two atoms: the localized one (1) and nonlocalized (2) in a standing spherical light wave, which reflects their trajectories in the laser field (the calculation was done by using (5.11)–(5.13), [Balykin et al., 1989c]). The atoms spaced of a distance d $= \lambda/2$ apart move parallel to each other and enter the field at a point near the minimum of two adjacent potential wells. The laser field parameter are as

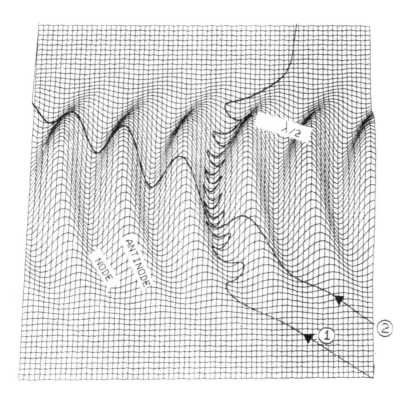

Fig. 28 Atomic potential energy in the spherical standing light-wave for a localized atoms (1) and nonlocalized one (2). The curves reflect the trajectories of localized and nonlocalized atoms [Balykin et al., 1989c].

follows: the laser power $P = 0.11$ W, the frequency detuning $\Omega = 400$ MHz, and the radius of curvature of the wave front is $R = 2$ m. The atoms differ in longitudinal velocity: the localized atoms moving with a velocity $v = 500$ m/s and nonlocalized atom moving with a velocity $v = 1200$ m/s. The atom moving with a larger velocity will have a transverse kinetic energy larger than the height of potential well and it will move free.

Figure 29 shows the influence of the impulse diffusion on the motion of atoms. The transverse kinetic energy of atoms at the entrance in the laser field is only a little less than the height of potential well. The impulse diffusion increases the kinetic energy and at certain point of standing wave it becomes larger the potential energy, which prevents atoms from being localized.

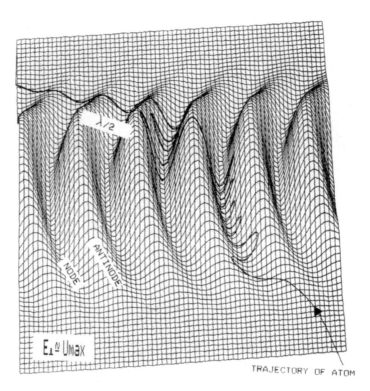

Fig. 29 The influence of the momentum diffusion on the motion of atoms. The transverse kinetic energy of atoms at the entrance in laser field only a little less than the height of potential well. The diffusion increases the kinetic energy and at a certain point of standing wave it became larger than the potential energy, which prevent atoms from being localized.

It is possible to improve stability of channeling of atoms by increasing the interaction time so that the atom could experience a velocity dependent gradient force. Such "dissipative" channeling was analyzed in [Kazantsev et al., 1985; Chen et al., 1993].

Figure 30 shows the change in transverse velocity undergone by localized atoms during their flight through the spherical standing wave, calculated by means of eq. (5.12). The atomic motion calculations were made both with and without taking the friction force into consideration. The friction force changes the amplitude of oscillation of a localized atoms but doesn't change considerably the character of motion of nonlocalized atoms.

We have considered the motion of atoms in a "mean" potential. This potential is appropriate for an atom in a standing wave when the spontaneous emission rate is

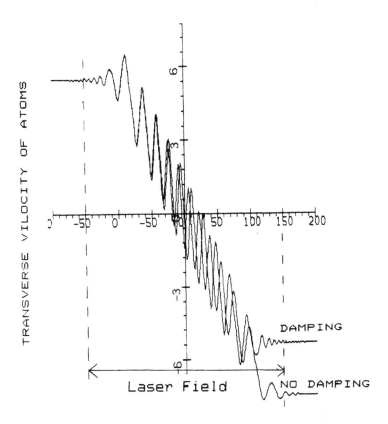

Fig. 30 The change in transverse velocity undergone by localized atoms during their flight through the spherical standing wave.

sufficiently high that the atom does not move a significant fraction of a wavelength between spontaneous emission events. When the rate of spontaneous decays is very small compared to the atom oscillations, the atomic motion is periodic in one of the dressed potential with interruptions caused by spontaneous emissions.

The consideration of channeling (both in a mean potential and in a particular dressed one) given above treated the atomic motion classically. When the transverse momentum of atom is on the order of $\hbar k$, its motion can no longer be considered classical. The quantization of an atomic motion in a standing wave has been recently observed with cooled and trapped atoms [Jessen et al., 1992; Verkerk et al., 1992; Hemmerich et al., 1993].

5.2 EXPERIMENTAL REALIZATION OF THE CHANNELING OF ATOMS

At present three different experimental methods were used to investigate the channeling of atoms in a standing wave. Prentiss and Ezekiel [1986] detected an increase of atomic concentration in the vicinity of the nodes or loops of the wave by measuring the fluorescence line shape of a beam of sodium atoms crossing a plane light wave at right angles. The detected asymmetry in the fluorescence line shape they explained as being due to the action on the atoms with the gradient force, coursing this concentration.

Salomon et al. [1987] used the absorption of an additional weakly resonant wave to determine the atomic density distribution in a standing wave. The density of atoms was found to increase near the nodes or loops of the standing wave, depending on whether the light frequency detuning was positive or negative with respect to the atomic transition frequency.

Balykin et al. [1988b; 1989a] demonstrated one dimensional localization and channeling of atoms in a spherical standing wave by observing a deviation of atomic beam from its original propagation direction.

Recently, Chen et al. [1993] have investigated the effect of the momentum diffusion on the channeling of atoms in an intense standing wave.

Hemmerich et al. [1991] have carried the experiment to investigate the channeling of atoms in two-dimensional standing wave. In two-dimensional case the relative time-phase difference between waves became an impotent parameter of motion of atoms. They have observed as channeling of atoms or their deflection.

Here we discuss the experimental realization of localization and channeling of atoms in (Fig. 31) a) plane standing wave by absorption method [Salomon et al., 1987] and in b) a spherical standing wave by deflection method [Balykin et al., 1988b; 1989a].

5.2.1 Plane wave

When atomic beam crosses standing wave and atoms have a low enough kinetic energy, they are guided into the channels where they move along the channel and oscillate in the transverse direction. To detect this channeling [Solomon et al., 1987] the atoms have been chosen themselves as probes of their position. Because of spatially varying of the laser field the light shift depends on the position of atom in a standing wave: for atoms at a node there is no light shift, elsewhere the absorption line is shifted by:

$$\delta\omega(r) = -\Omega\left\{\left[1 + \omega_R^2(\rho)/\Omega^2\right]^{1/2} - 1\right\} \tag{5.18}$$

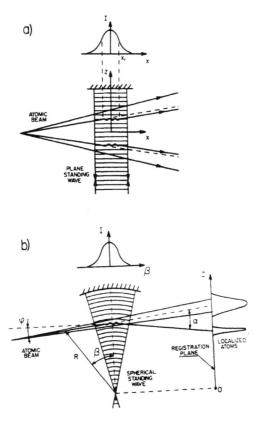

Fig. 31 Two scheme of channeling of atom in a standing wave. (a) Localization of atoms in a plane standing wave. Top: intensity distribution along the transverse coordinate of the beam forming the standing wave. x_1 is the distance from the laser beam center at which atomic localization occurs. (b) Channeling of atoms in a spherical standing wave.

Figure 32 (c,d) show the calculated absorption spectrum of atoms in a standing wave. The spectrum was obtained by adding the contribution of the atoms in the various locations in a standing wave. The curve (c) corresponds to a uniform spatial distribution (no channeling) for a blue detuning in a standing wave, the curve (d) is calculated with a periodic spatial distribution of atoms channelled near the nodes. The modification induced by channeling clear appears in the second case. Peak N, corresponding to the atoms near the nodes, is enhanced while peak A, corresponding to the antinodes, is, weakened.

The experiment has been performed with Cs atomic beam. The atoms were prepared as two-level atoms by a method of optical pumping. The intensive standing wave (the

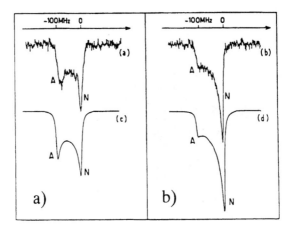

Fig. 32   (a) The calculated absorption spectrum of atoms in a standing wave: 1 – the uniform atomic distribution in standing wave; 2 — the channeling of atoms. (b). The experimental absorption spectrum of atoms in a standing wave [Salomon et al., 1987].

Rabi frequency −210 MHz, the detuning ±150 MHz) has irradiated the atomic beam at right angle. The weak probe beam was also orthogonal to the atomic beam and travel through the central part of the standing wave. The fractional absorption of atoms was only 10^{-5}. The modulation of absorption and using the synchronous detection allows to measure the absorption of as few as ten atoms in the 2 mm^3 observation volume.

The curve (a) was measured with a tilted standing wave: there is no channeling. The curve (b) – with an orthogonal standing wave: the channeling should be in this case near the nodes. The maximum height of the potential hill was 2 mK, which corresponds to the maximum trapping velocity equal to 50 cm/s. Both experimental curves fit with the calculated ones. From the experimental absorption spectra have been deduced corresponding spatial distributions of atoms which have shown the concentration of atoms near the nodes of standing wave.

5.2.2 Spherical wave

The basic idea of the experiment with a spherical wave is schematically illustrated in Fig. 31b. The atomic beam traverses the spherical standing wave at a point far from the beam waist. In the polar system of coordinates R, β the effective potential has the form:

$$U(z) = U_0 \cos^2 kz + F R \qquad (5.19)$$

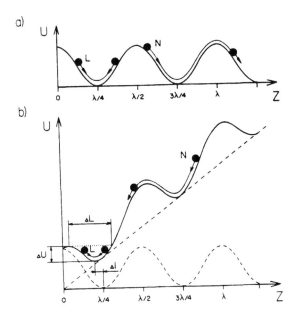

Fig. 33 Atomic localization in a) plane and b) spherical standing waves. The circular motion in spherical wave of atoms gives rise to an effective centrifugal force, which causes the potential to change as shown by the solid line.

where the first term is the atomic potential in the laser field, the second one — an inertial potential. The presence of the effective field has the following effects. First, it reduces the depth of the atomic potential well (see Fig. 33) to

$$\Delta U = U_0 \left[1 - (F/U_0 k)^2\right]^{1/2} + (F/k)\,[\arcsin\,(F/U_0 k) - \pi/2] \qquad (5.20)$$

where $U_0 k$ is the maximum gradient force in the standing wave in the absence of inertial force. Second, it reduces the size of the potential well. Third, it shifts the position of the well bottom. Finally, what is most essential is the fact that the inertial field gives rise to the force averaged over the standing wave period. This force accelerates the nonlocalized atoms and hence causes their spatial separation from the localized atom. This makes it possible, first, to measure the atomic localization effect itself by observing the spatial separation of the atoms, and second, to isolate cold (localized) atoms from nonlocalized ones.

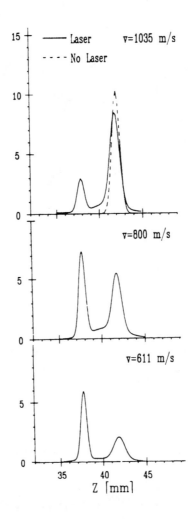

Fig. 34 The experimental results of the localization of sodium atoms in the spherical standing wave. The curves are the spatial profiles of the atomic beam in the measured region. The atomic velocity value: v=611, 800, and 1035 m/sec. After interacting with standing wave the atomic beam gets split into two. The left peak corresponds to localized atoms and the right to nonlocalized ones [Balykin et al., 1989c].

In the experiment the spherical standing wave was produced by focusing a laser beam into the center of curvature of a spherical mirror and by reflecting the beam back by this mirror. The standing wave diameter at the point of interaction with atomic beam was 0.6 mm and the wave-front radius -40 mm. The laser intensity in the atom-field interaction region was 20 W/cm^2. The atomic beam profiles was measured by means of probe laser radiation tuned to resonance with atomic transition. For this purpose, the narrow probe beam transversed the atomic beam at small angle and to scan it in space at the certain point from the spherical wave.

Figure 34 presents the experimental results of the channeling of sodium atoms in the spherical standing wave. The curves are the spatial profiles of the atomic beam in the measured region. The measurement were done for three atomic velocities: v = 611, 800 and 1035 m/sec. It can be seen from the figure that after interacting with the standing wave, the atomic beam gets split into two beams. The left peak corresponds to localized atoms and the right to nonlocalized atoms. This conclusion follows from comparison with the calculated curve. The distance between the peaks in experimental and theory agrees accurately enough. This distance is determined by wave-front curvature and the size of the laser beam in the atom interaction region.

5.3 COLLIMATION OF ATOMIC BEAM THROUGH THEIR CHANNELING

As we have seen in previous subsection that an atom can be localized in spherical standing wave if 1) its trajectory is a tangent to the wave front of a field at the entrance point, 2) the gradient force exceeds the centrifugal force and 3) the intensity is growing in the direction of atomic motion. During the localization of atoms the transformation of longitudinal velocity into transverse velocity takes place. If the law of increasing of intensity of light is different at an entrance edge of a standing wave compared to an exit edge then a transverse velocity of atoms to the exit could be less than initial one, i.e. an effective transverse cooling of atoms could take place. If the entrance edge of the laser beam has the Gaussian profile and the intensity to the exit edge drops sharply than the relative transverse cooling of the atomic beam after crossing such standing wave is defined by relation [Balykin et al., 1990]:

$$T_f/T_{in} \cong (4e^2/\pi)(R\lambda/\rho^2) \qquad (5.21)$$

where T_f and T_{in} are the transverse temperature of atomic beam at the exit and the entrance of standing wave. The ratio (5.21) contains only the geometrical parameters of the standing wave at the point of intersection with the atomic beam: R is the radius of curvature of the wave at the point of intersection with the laser beam, ρ is the radius of laser beam.

The transverse cooling of atoms through the channeling in the truncated spherical standing wave was observed in [Balykin et al., 1990]. The standing truncated wave was formed by a screening of the beam of a single-frequency CW laser by half with

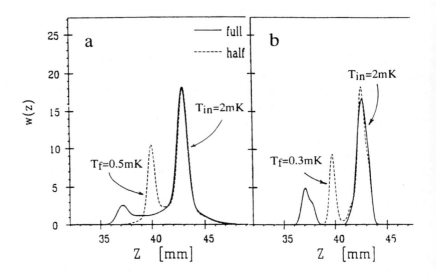

Fig. 35 The experimental results of transverse cooling through a channeling. The solid curve corresponds to the case of the whole spherical wave and the dashed one, to that of the truncated wave. The peak 1 is produced by the atoms localized in the whole spherical wave; the peak 2 is formed by the atoms localized in the truncated wave, and the peak 3 — by nonlocalized atoms (a — experiment, b — theory) [Balykin et al., 1990].

a sharp edge. The transverse cooling was detected by measuring the spatial profiles of atomic beam after their intersection with the standing wave. Figure 35 presents the results of such a measurement. The solid curve corresponds to the case of the whole spherical wave and the dashed one, to that of the truncated wave. The left peak is produced by the atoms localized in the whole spherical wave; the middle peak is formed by the atoms localized in the truncated wave, and the right peak by nonlocalized atoms. The width of peak formed by localized atoms in truncated wave, is more narrow than the width of peak of nonlocalized atoms. The transverse temperature was reduced from $k_B T_{in} = 8.5\ \hbar\gamma$ to $k_B T_f = 2.1\ \hbar\gamma$. The numerical modeling of this experiment gave the value for the final temperature $T_f = 1.3\ \hbar\gamma$, that was in good agreement with the experimental value.

Another scheme of a cooling of atoms through their channeling is based on the adiabatic lowering the standing wave intensity. Once the atoms are channeled, their kinetic energy may be reduced by adiabatically lowering the standing wave intensity at a rate that is slow in comparison with the atom's oscillation period. It was demonstrated that channeled atoms can be cooled in this scheme to a single-photon recoil limit [Chen et al., 1992].

5.4 BICHROMATIC LASER FIELD

As we have seen in the previous section in a plane intensive standing wave the gradient force can be a considerably strong one and can be used for localization of atoms on the spatial scale of $\lambda/2$. Application of this force to the manipulation of the atomic motion is restricted by the fact that this force oscillates in the scale $\lambda/2$ and it cannot effectively act on the macroscopic scale $\gg \lambda/2$. This restriction was overcome by proposal of Kazantsev and Krasnov [1987; 1989] by using the gradient force in a standing wave formed by bichromatic standing wave. Later this idea was developed to the case of three-level atom in bichromatic wave and three-level system in monochromatic wave (Javanainen, 1990; Grimm. et al., 1991). The force that appears in such a field is constant in sign over macroscopic scales that considerably exceed the wavelength of light.

The main idea of rectification of the gradient force in a bichromatic standing wave is following. The behavior of the gradient force in monochromatic standing wave and is governed by three factors: the saturation parameter G, the detuning of laser frequency from the atomic absorption frequency Ω and the gradient of light intensity $\nabla G (G = I/I_s)$:

$$\mathbf{F} = -(1/2)\hbar\Omega\left[1/\left(1 + G + \Omega^2/\gamma^2\right)\right]\nabla G \qquad (5.22)$$

The sign of the gradient force in a monochromatic standing wave is determined by the sign of the gradient of light intensity and by the sign of the detuning of the laser frequency. The sign of the gradient intensity oscillates on the wavelength scale. If we could manage to find the physical process which changes also the sign of detuning on the same spatial scale, the gradient force would be constant in the sign. Such a process, pointed out by Kazantsev and Krasnov [1987], is the AC stark shift of atomic transition from an additional second laser field. The first and the second laser fields should fulfill the following conditions: the second field is strongly detuned from atomic resonance, so the corresponding detuning Ω_2 exceeds the detuning of the first field Ω_1 and Rabi frequencies of both fields ω_{R1}, ω_{R2}:

$$\Omega_2 \gg \Omega_1,\ \omega_{R1},\ \omega_{R2},\ \gamma \qquad (5.23)$$

At such conditions the rectified force and its potential in the field of two plane standing waves $E_1(x) = E_1 \cos kx$, $E_2(x) = E_2 \cos[(k + \delta k)x + f]$, $(\delta k \ll k)$, is

$$\langle F_{gr}\rangle_\lambda = -\hbar k\gamma \left[\left(\omega_{R1}^4/\Omega_1^3\gamma\right)\left(\omega_{R2}^2/\Omega_1\Omega_2\right)\right]\sin 2(k_1 - k_2)x \qquad (5.24)$$

The average force (5.24) of the same order as the gradient force in monochromatic laser wave $(\langle F_{gr}\rangle_\lambda \approx \hbar k\Omega_1)$ and it oscillates with period $1 = \pi/\delta k$, which can be

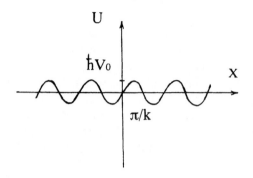

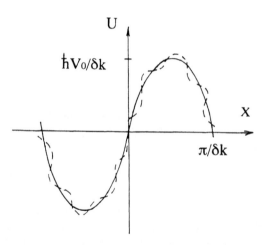

Fig. 36 a) The potential of gradient force in monochromatic standing wave; b) the average potential (solid curve) and unaverage potential (the dashed line) in bichromatic light field [Kazantsev and Krasnov, 1987].

considerably larger than the wavelength of light. The depth of the potential wells of the rectified force, $U \cong \hbar k \Omega_1 / \delta k$, becomes larger than the one frequency potential $\hbar \Omega_1$ by the factor $k/\delta k$. Figure 36 illustrates schematically the potential of the gradient force of the monochromatic standing wave (a) and the average potential (solid curve) and unaverage potential (the dashed line) in bichromatic light field (b). The unaverage potential contains also the small scale oscillations with amplitude and period of the wells in monochromatic field. The average potential in bichromatic field can be

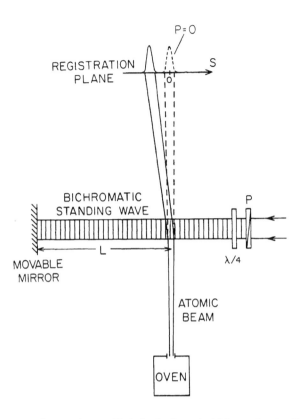

Fig. 37 Set up to observe the rectified dipole force in bichromatic standing wave. P —
polarizer; λ/4 — quarter-wave plate [Grimm et al., 1990].

considerably larger than in monochromatic field. For example, for Na atoms in the
laser field of 1 W/cm^2 and $\delta k/k = 10^{-4}$, the depth of potential is 20 K.

The first test of the rectification of gradient force was done in [Grimm et al.,
1990]. The experiment schematically shown in Fig. 37. An atomic beam crosses
the bichromatic standing wave which is formed by reflection from the plane mirror
two frequency laser beam. The distance between the frequencies is 1.45 GHz, a
corresponding period of rectified force is $l = 10.3$ cm. The high frequency is tuned
close to the atomic transition of Na atom ($\Omega_1 = +140$ MHz). The other frequency
is strongly detuned from atomic resonance ($\Omega_2 = -1.31$ GHz). In the experiment
the deflection of atomic beam was measured at several different intersection points of
atoms with a laser beam, which correspond to the bottom of potential well and to the
left and the right slopes of the well. Fig 38 shows the results of such measurements.

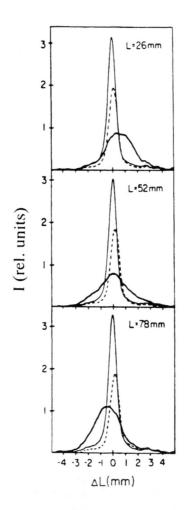

Fig. 38 The test of rectification of gradient force in bichromatic laser field. Atomic beam profiles in the measured region for three different crossing of atomic beam and the bichromatic laser field: a) and c) corresponds to the left and the right sides of potential well; b) to the bottom of well. The dashed curve is an initial profile [Grimm et al., 1990].

When atoms cross the well at the bottom there is no deflection ((b) curve). The crossing of the well at the slopes gives the deflection of atoms ((a) and (c) curves) which corresponds to the action of rectified force. The average maximum force observed in experiment is $\langle F \rangle_\lambda = 3.4 \, \hbar k \gamma$.

In this Section we have considered only one-dimensional monochromatic and bichromatic standing wave. In the two dimensional case the situation becomes more complicated [Kazantsev and Krasnov 1987; 1989; Hemmerich et al., 1991]. By changing the directions of propagation of the interfering fields, it is possible to create either potential or a vortex rectified force. As in the case of one dimensional standing wave a potential rectified force creates the deep potential wells. The atomic motion under the potential and vortex rectified forces turns out to be either rotational or a random walk.

6. REFLECTION OF ATOMS BY LIGHT

6.1 Laser field and atomic potential in a evanescent wave

We consider the reflection of atoms in the field

$$E = E_0(x) \exp[i\theta(r)] \tag{6.1}$$

in which the amplitude of the light $E_0(x)$ is changed adiabatically slow in comparison with relaxation of the internal atomic motion. In this case the light pressure force in the field is usually split into parts:

$$F = F_{LP} + F_{GR} \tag{6.2}$$

The first part of this force was called a light pressure force (see Section 2) and it is determined by a gradient of a phase of the laser field and can be represented as:

$$F = \chi'' I \nabla \theta \tag{6.3}$$

where $\langle \mu \rangle = \chi E$ and $\chi = \chi' + \chi''$ is the susceptibility and I the intensity of the light. This force was used for collimation and focusing of atomic beams.

The second part of force (6.2) is related to the intensity gradient of laser field:

$$F_{GR} = \chi' \nabla I \tag{6.4}$$

This force is called a gradient force and can be used for example for deep focusing of an atomic beam.

By choosing the parameters of the laser field (a detuning, an intensity, a size of beam,...) it is possible to realize an interaction of atom with field when:

$$F_{LP} \ll F_{GR} \tag{6.5}$$

and $F \approx F_{GR}$: the radiation pressure force (6.2) will be a potential force. The maximum value of this force can be very large. At the size of laser field $\rho = \lambda$, and a detuning of order of the Rabi frequency $\Omega \approx \omega_R$, the gradient force is:

$$F_{GR} \approx (\hbar\Omega/4I)\nabla I \cong \hbar\omega_R/\lambda \approx F_{LP}(\omega_R/\gamma) \tag{6.6}$$

As can be seen from (6.4) the gradient force is directed along the gradient of the laser field and the sign of this force is determined by the sign of the detuning. At a positive detuning the gradient force expels an atom from the laser field. These properties of the gradient force can be used for reflection of atoms from a laser field.

83

There are several configurations of laser field which can be used for the reflection of atoms.

Let us first consider the scheme suggested by Cook and Hill [1982]. When a plane traveling light wave is totally reflected internally at the surface of a dielectric in a vacuum, a thin wave is generated on the surface. By an application of Fresnel's reflection formulas [Born and Wolf, 1984], it can be found the intensity of evanescent wave:

$$I(x) = qI_0 \exp(-x/x_0) \qquad (6.7)$$

where

$$q = 16n^3 / \left[(n^2 - 1)(n + 1)^2 \cos^2 \phi \right] \qquad (6.8)$$

$$x_0 = \lambda / \left[4\pi \left(n^2 \sin^2 \phi - 1 \right)^{1/2} \right] \qquad (6.9)$$

and

$$I_0 = I_0^0 \exp\left(-\rho^2/\rho_0^2\right) \qquad (6.10)$$

is the intensity of the laser radiation that is incident on the surface of a dielectric in a vacuum, $I_0^0 = P/\pi\rho_0^2$; P is the laser radiation power; ρ_0 is the radius of the laser beam sport on the dielectric surface; n is the refractive index of the dielectric; ϕ is the angle of incidence of laser beam; x is the distance from the dielectric surface. The evanescent wave propagates parallel to the surface and in the direction, normal to the surface, this wave decays at the distance of $x_0 \cong \lambda/2$, where λ is the length of light wave. The characteristic depth of penetration of the field into the vacuum is of the order of wavelength.

If we put the atom in such evanescent wave, the atom in this evanescent wave will experience a radiation force due to momentum transfer from the wave. For two level atom, the radiation force has a component parallel to the surface (the light pressure force) and a component normal to the surface (the gradient force). The maximum value of gradient force is reached at the detuning of the order of the Rabi frequency. At such detuning the light pressure force is considerably smaller than the gradient force and it could be neglected in the interaction of atom with evanescent wave. The expression for the gradient force is:

$$F(y) = \kappa\hbar(\Omega - k_1 v_x) \left\{ G(y) / \left[1 + G(y) + (\Omega - kv_x)^2/\gamma^2 \right] \right\} \qquad (6.11)$$

where $\kappa = k\left(\sin^2\phi - n^{-2}\right)^{1/2}$ and $k_1 = k\sin\phi$ are the imaginary and real components of the wave vector $k = \lambda/2\pi$, v_x is the x component of the atomic velocity along the surface, $G = I/I_S$ is the saturation parameter, $I(y)$ being the laser intensity and I_S the transition saturation intensity.

It is this surface wave that can be served as an atomic mirror for an atom running into it: with a positive detuning, the gradient force expels atoms out of the field.

It can be shown [Cook and Hill, 1982; Ol'shanii et al., 1992] from a consideration of the motion of atoms under the action of the gradient force (6.11) (and when the amplitude of the light is changed adiabatically slow in comparison with relaxation of the internal atomic motion) that the angle of reflection of the atom is equal to the angle of incidence, which means that we have here a specular reflection of the atom.

The reflection of atoms at "the mirror" may be also considered as a result of its being expelled out of a potential field whose energy is

$$U(y) = 1/2\hbar(\Omega - k_1 v_x)\ln\left\{1 + G(y)/\left[1 + (\Omega - k_1 v_x)^2/\gamma^2\right]\right\} \qquad (6.12)$$

With the transverse atomic velocity component $v_\perp$ being great enough ($v_\perp > v_{max} = [2U(0)/M]^{1/2}$, M being the atomic mass), the atoms reach the surface. In this case, they either adhere to the surface or are reflected from the surface. Usually the reflection of atoms by the surface is diffusive. Specular reflection can occur when the de Broglie wavelength of atom divided by the grazing angle is large then the local surface roughness [Ramsey, 1956; Anderson et al., 1986]. The atomic mirror can effectively reflect the atoms whose maximum velocity component to the mirror surface is given by:

$$v_{max}^2 = 2U(0)/M \approx v_r v_\gamma G(0)q(\gamma/\Omega) \qquad (6.13)$$

For typical atomic parameters (e.g., for sodium atom) $M = 4 * 10^{-23}g$, $v = 6 * 10^4$ cm/s, and $\gamma/2\pi = 5$ MHz, and laser parameters $k_1 = 10^5$ cm^{-1}, $G(0) = 10^5$, and the laser detuning 2.6 GHz, we have $v_{max} = 430$ cm/s. Accordingly, for thermal atomic velocities, the maximum grazing angle is $7 * 10^{-3}$ rad.

In the reflection of atoms by evanescent wave the amplitude of the light field can be changed not necessarily adiabatically slow in comparison with relaxation of internal atomic motion. In this case the expression (6.11) for light pressure force acting on the atom are only zeroth-order term of expansion of force into powers of inverse interaction time, [Ol'shanii et al., 1992]. The next term in expansion gives rise to a dissipative part in the gradient force. Such nonadiabaticity can happen if the time of interaction of atom of this field is comparable with γ^{-1}. In this case the specular character of reflection of atoms can be disturbed.

Another approach to describe the reflection of atoms by evanescent wave is the dressed state picture (see Section 5). For two level atom its dressed state energy is

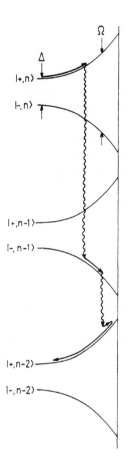

Fig. 39   Dressed state picture of an atom reflection in evanescent wave [Seifert et al., 1993].

given by (5.1) and (5.2) where Ω_R is the spatially dependent Rabi frequency of an evanescent wave. An atom in a dressed state undergoes force $-F$ equals to minus the gradient of the energy of this state. The spatial dependence of the evanescent wave is shown schematically in Fig. 39. For positive detuning an incoming atom initially in the ground state $|g>$, evolves addiabatically in the state $|+, n>$. In this state atom experiences a repulsive potential: the atom climbs the potential hill and decelerates. If the kinetic energy of atom is small compared to the maximum value ot the potential it will be stopped at the classical turning points and the direction of motion is reversed; it will be completely reflected.

If a spontanenus emission occurs the atom can go to a dressed state $|+, n - 1 >$ or $|-, n >$. In the state of type (+) atom sees the same repulsive potential as before. In the state of type (−) it sees an attractive potential. Because of the random character of spontaneous emission, the instantaneous force experienced by atom switches back and forth between two opposite values, when the atom is moving in an evanescent wave. The mean dipole force acting upon an atom in an evanescent wave fluctuates around their mean value, and these fluctuations produce a diffusion of atomic momentum and this leads to the departure from ideal reflection.

6.2 REFLECTION OF ATOMS FROM A GAUSSIAN LASER BEAM

An atomic mirror can be formed also by a focused elliptical laser beam, in which atoms intersect near the waist of the laser beam [Balykin, 1989d]. The intensity distribution of laser beam can be expressed in this case as:

$$I(x, y, z) = I_0(q/q_0)(p/p_0) \exp\left[-\left(x^2/q_0^2 + y^2/p_0^2\right)\right] \qquad (6.14)$$

where q, p, q_0, p_0 denote the laser beams waist along x- and y-axes, I_0, P is the laser radiation intensity at waist and laser power. The atom is moving in the x-z plane. The atomic potential energy U in this laser field is expressed by (2.23).

When the atom with kinetic energy $E < U$ impinges upon the potential barrier of such laser beam, it will be reflected by this barrier. If the energy of atom is higher than the height of potential barrier the atom will penetrate the barrier. At the atomic energy $E \cong U$, there is a noticeable probability that this atom will tunnel through the barrier. In this case such laser beam can be considered as a semitransparent atomic mirror or a coherent beam splitter. The tunneling will be observed only for atoms whose kinetic E energy satisfied the following relation:

$$[2M(U - E)1/2]/l \cong \hbar \qquad (6.15)$$

where l is a width of a potential barrier (in the considered case — the diameter of laser beam), M is a mass of an atom.

The transmission coefficient Θ of a potential barrier whose shape depends quadratically on a coordinate near the top can be expressed as:

$$\Theta = [1 + \exp(-\zeta)]^{-1} \qquad (6.16a)$$

where

$$\zeta = 2\pi [(E - U_0)/\hbar] \left(Mq_0^2/2U_0\right) \qquad (6.16b)$$

The parameter ζ, can be expressed also in terms of atomic velocity as

$$\zeta = (4\pi^2 q_0/\lambda)(v - v_{thr})/v_r \qquad (6.17)$$

where $v_{thr} = (2U_0/M)^{1/2}$ is the threshold velocity of the atom whose energy is equal to the potential barrier height. To observe the tunneling of atoms its initial velocity distribution should be very narrow. For a laser beam waist $q \sim \lambda$ the velocity width of atomic beam should be in the range: $\delta v \leq (1/8\pi^2)(v_r^2/v)$. For sodium decelerated atomic beam with mean velocity $\langle v \rangle = 10^2$ cm/s the velocity width should be $\delta v = 10^{-3}$ cm/s. Such atomic mirror could also be used as velocity selector for atomic beam.

6.3 REFLECTION BY SURFACE PLASMON WAVE

The surface plasmon waves are the surface electromagnetic modes that travel along a metal — dielectric interface as bound nonradiative waves with their field amplitude decaying exponentially perpendicular to the interface (Raether, 1988). Surface plasmons are usually excited by coupling them to an evanescent wave at a dielectric surface. A plasmon wave as an atom mirror can be prepared in the following way. On a glass surface a thin metallic layer (with a typical thickness of 70 nm) is evaporated. A laser field is totally internally reflected from the glass surface. By varying the angle of incidence of the light beam, the wave vector of the evanescent wave can be varied. When the wave vector coincides with the surface plasmon vector, most of the laser photons are converted into the surface plasmons. An attractive feature of a plasmon wave as an atom mirror is a large field enhancement of the initial laser beam intensity. The field enhancement ratio ϑ is determined as a ratio between the maximum intensity of the evanescent field with plasmon excitation and the field intensity of the evanescent wave on the bare dielectric surface. The maximum electric field enhancement is (Raether, 1988):

$$\vartheta_{max} = (1/\varepsilon_2) \left(2 |\varepsilon_1'|^2 / \varepsilon_1'' \right) \left[a/ \left(1 + |\varepsilon_1'| \right) \right] \qquad (6.18)$$

where $\varepsilon_1 = \varepsilon_1' + i\varepsilon_1''$, ε_2, ε_0, are the dielectric constant of the metal, the vacuum, and the dielectric, $a^2 = |\varepsilon_1'|(\varepsilon_0 - 1) - \varepsilon_0$. If the dielectric is quartz ($\varepsilon_0 = 2.2$), the metal layer is $Ag(\varepsilon_1' = -17.5, \varepsilon_1'' = 0.5)$ and the wavelength $\lambda = 700$ nm, then the enhancement factor can be $\vartheta \sim 100$. The resonance condition of plasmon excitation depends on the thickness of the metal layer that results in a thickness dependence enhancement of plasmon wave. This effect could cause a strong fluctuation of the field intensity.

A simple evanescent wave allows one to create a high homogeneous field along the surface of an atomic mirror, without a limitation on the intensity of the field in an evanescent wave. The size of an atomic mirror is usually restricted by the laser intensity required by the reflection. The main motivation to use plasmon wave as an atomic mirror is the possibility of reaching a high field intensity with a low laser power. There are several drawbacks to this scheme using plasmon waves as an atom mirror. The metal layer strongly absorbs light which leads to the distraction of the layer and this process limits the intensity. The intrinsic roughness of the interface produces a strong variation of the local field intensity of the plasmon wave and couples out some of the intensity to the vacuum side. Both effects lead to a broading of atomic beam during the reflection.

6.4 EXPERIMENTS

6.4.1 Simple evanescent wave and plasmon wave

The first observation of a specular reflection of atoms has been done in [Balykin et al., 1987d]. Atomic mirror was a parallel-face plate of fused quartz 0.4 mm thick and 25 mm long into which the laser beam enters through a beveled side (Fig. 40). A multiple total internal reflection of the laser was used to increase the surface area of the atomic mirror. The power of the laser beam was 650 mW and its diameter 0.4 mm. Figure 41 shows the arrangement of the reference, incident and reflected atomic beams. In the experiment the beam of sodium atoms was used. When the atomic mirror is parallel to the atomic beam axis so that atoms fly past it, the profile of the atomic beam in the detection region shows only the reference and incident beams. If the mirror is tilted so it cuts off the atomic beam, the latter is observed to be reflected. As one increases the tilt angle still further, the angle of reflection increases as well, but the number of atoms undergoing mirror reflection decreases because some atoms reach the mirror surface and suffer diffusive reflection. The maximum angle of reflection observed in this experiment was about 0.4°. An important parameter of any mirror is its reflectivity. At small incident angle, the reflectivity of atomic mirror was observed close to 100%.

In another experiment [Kasevich et al., 1990] the specular reflection of atoms was observed by dropping a sample of laser cooled atoms on the evanescent wave. The sample of cold atoms was prepared in the following way. Na atoms were loaded into opto-magnetic trap by slowing down a thermal atomic beam with counter propagating frequency-chirped laser beam. The trap was formed by three mutually orthogonal pairs of counter propagating, circular polarized laser beams intersecting in zero field region of a spherical quadrupole magnetic field. After loading of atoms during of 0.5 s, the field coils were turned off, leaving a small residual magnetic field in order to cool the atom further in optical molasses. The final stage of cooling had been done by

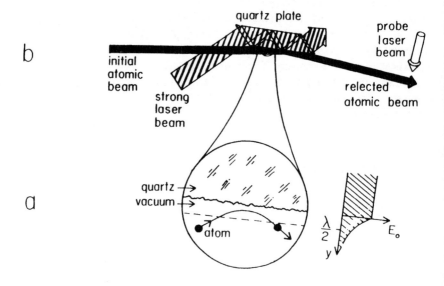

Fig. 40 (a) Principle of mirror reflection of atoms by nonuniform laser field of surface wave. (b) Schematic diagram of experimental setup.

the gradual extinction of the light. The final temperature of the atoms (the cloud of 3 mm diameter) was T $\approx$ 25 μK. The evanescent wave was created by total internal reflection of laser beam of 8.9 W/cm^2 intensity at the detuning 400 MHz. The atoms were dropped from a height of 2 cm and two bounces were registered at the initial trapping region. After the first bounce about 0.3% of the atoms were observed in the trapping region. The number of atoms after the second bounce was 10% of those observed after first bounce. The main losses for atoms was the ballistic expansion of the sample of atoms having a certain initial spatial and velocity extents.

Recently several groups have reported the experimental realizations of an atomic mirror which were based upon the dipole force exerted by a surface plasmon wave. The specular reflection of a supersonic metastable argon [Seifert et al., 1993] and metastable neon [Feron et al., 1993] were demonstrated with plasmon wave at metal-vacuum interface. The enhancement of laser field by plasmons increased the maximum reflected angle by a factor of $\sim$ 2 $\div$ 3.

In another paper [Esslinger et al., 1992] an intensity enhancement of nearly two orders of magnitudes in evanescent wave permitted the use of a low power diode laser for reflection of Rb atoms at $\sim$ 2 mrad.

An atomic mirror could be a device of great practical importance in its applications such as a recombiner in atom interferometer, for a deep focusing (as concave mirror) and in storage of atoms in an atomic cavity. In all of these potential applications it is essential that the atom mirror be an "ideal mirror" in that: a) a reflection is a specular reflection, b) a phase shift introduced through a reflection is calculable ("coherent" reflection).

There are several processes at the reflection of atom by light which could lead to non ideal properties of the mirror. First, spontaneous emission of photons by atom during its interaction with light, second, a spatial variation of the laser intensity due to the inhomogeniety of laser beam and the roughness of dielectric surfaces. In [Seifert et al., 1993] it was investigated that the influence of both processes on the reflection of atoms by two types of evanescent waves: a) a simple evanescent wave and b) an evanescent wave enhanced by surface plasmons excited in a thin metallic layer. It was realized the scheme of reflection of atoms based on the use of an "open" transition in which a spontaneous emission was almost completely eliminated: 90% of atoms were reflected coherently.

6.4.2 Quantum State Selective Reflection of Atom

One of the remarkable properties of an atomic mirror is its ability to reflect atoms in a certain quantum state. That the atomic mirror is quantum state selective follows from the character of the relationship between the gradient force and the detuning. When the detuning is positive, the gradient force repels an atom from the surface and thus the specular reflection takes place. With negative detuning, the force attracts an atom to the surface, and so diffusive reflection is observed.

Let an atom (or molecule) have several sublevels in the ground state. For the atoms, there may be, for example, fine and hyperfine ground-state sublevels (and for molecules, the vibration-rotational sublevel of the ground state). Atoms (molecules) on a sublevel for which the transition frequency to an excited state is lower than the laser frequency are reflected from atomic mirror. Thus, if a beam of atoms (molecules) distributed among several ground state sublevels is incident upon the mirror, the reflected beam will contain only atoms in one and the same quantum state.

The quantum state selective reflection was studied with sodium atoms [Balykin et al., 1988c]. The ground state of sodium atom $3S_{1/2}$, is split, because of hyperfine interaction, into sublevels, one with the quantum number $F = 2$ and the other with $F = 1$, the distance between which is 1772 MHz. According to the statistical weights of these sublevels, 62.5% of sodium atoms in a thermal beam are on the sublevel with $F = 2$ and 37.5% on that with $F = 1$. If the laser frequency ω is selected so that the condition $\omega_1 < \omega < \omega_2$ (ω_1 and ω_2 being the frequency of the transitions from the sublevel with $F = 1$ and $F = 2$ to the exited state $3P_{3/2}$) is satisfied, the reflected beam will contain only the atoms in the quantum state with $F = 2$.

92   V.I. BALYKIN AND V.S. LETOKHOV

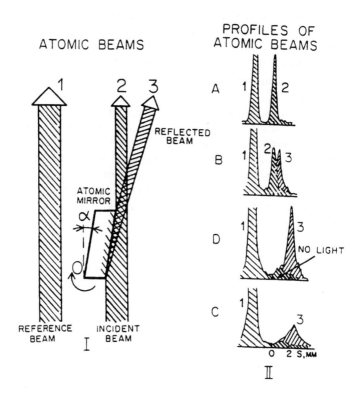

Fig. 41   Reflection of atomic beam by light. Atomic mirror is a parallel plate of fused quartz
into which the laser beam enters through a beveled side. A multiple total internal reflection of the
laser was used to increase the surface area of the atomic mirror. Figure shows the arrangement
of the incident and reflected atomic beams. When the atomic mirror is parallel to the atomic
beam axis so that atoms fly past it, the profile of atomic beam in the detection region shows
only the reverence and incident beams (A). If the mirror is tilted so it cuts off the atomic beam,
the latter is observed to be reflected (B, C, D). The dashed curve is the profile without laser
field [Balykin et al., 1987d]).

Figure 42 shows quantum-state-selective reflection of an atomic beam at atomic
mirror. When the detuning of frequency of laser field forming the atomic mirror is
positive to the transition F = 2, $S_{1/2} - 3P_{1/2}$, and negative to transition F = 1,
$S_{1/2} - 3P_{3/2}$ the absorption spectrum of the reflected atomic beam contains the atoms
only on the sublevel with F = 2 (Fig. 42(a)). The presence of a small number of atoms
at the sublevel F = 1 is explained by the scattering of sodium atoms on the residual
gas molecules. The subtraction of the signal due to scattering atoms (dashed line in

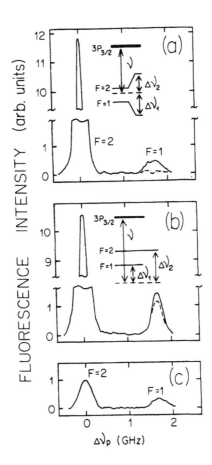

Fig. 42 Quantum-state-selective reflection of atomic beam at atomic mirror. (a) Absorption spectrum of reflected atomic beam at the detuning of intense laser field positive to the transition F=2–3P$_{3/2}$, and negative to transition F=1–3P$_{1/2}$. The reflected beam contains the atoms only on the sublevel with F=2. (b) The detuning of intense laser field is positive to the both transition. The reflected beam contains the atoms on the both sublevels. (c) Scattered atoms at P$_{las}$=0. The dashed line, the result of subtraction of background signal due to scattered atoms [Balykin et al., 1988c].

Fig. 42(a)) shows that there is no reflected atoms on the sublevels F = 1. At a positive detuning of intense laser field to both transition the reflected beam contains the atoms on both the sublevels, Fig. 42(b). The absorption spectrum of scattered atoms (at P_{las} = 0) contains the atoms on both sublevels, Fig. 42(c).

It was also measured the reflection selectivity of the atomic mirror which represents the ratio of the reflection coefficient of the atoms on the sublevels $F = 2$ and $F = 1$. The maximum experimentally observed reflection selectivity was about 100 and it was determined by the noise of the registration system.

As we have already mentioned that a quantum state selective reflection can be expected in the case of molecules as well. This will offer the unique possibility of preparing and spectroscopically analyzing beams of molecules in one and the same vibrational-rotational state which can be changed by varying of the laser frequency.

6.5 DIFFRACTION OF ATOM BY EVANESCENT WAVE

Diffraction of atoms by light has received considerable interest with regard to its practical importance for atomic interferometer (coherent beam splitter) and its intrinsic features. Pritchard and co-workers have demonstrated the diffraction of sodium atoms at normal incidence to a transmission grating consisting of an optical standing wave [Gould et al., 1986; Martin et al., 1988]. Hajnal and Opat [1989] proposed to combine reflection and diffraction of atoms by using a standing evanescent wave. The required optical wave field can be produced by totally internally reflecting a laser beam at the surface of a refractive medium and the retro-reflecting the light back along its original path. The evanescent field decreases exponentially in the direction to the surface and is modulated sinusoidally along the surface.

The result of diffraction of atom on the standing evanescent wave can be expected by considering the incident atomic beam as a plane de Broglie wave incident upon a periodic structure formed by standing evanescent wave. The diffraction angles of the reflected atomic beams are given by grating equation:

$$(\lambda/2)(\sin \Phi_{rm} - \sin \Phi_i) = m\lambda_B \qquad (6.19)$$

where Φ_{rm} and Φ_i are the reflected and incident angles of the atomic beams, λ_B is the de Broglie wavelength corresponding the atomic beam, m is the order of diffraction. The diffracted beams are separated by the angle:

$$\delta\Phi_{rm0} \cong 2(\lambda_{Br}/\lambda)/\Phi_{gl} \qquad (6.20)$$

where Φ_{gl} is a incident glancing angle to an evanescent wave. At small glancing angle the diffraction angle can be considerably larger (a factor of 100) than one in the diffraction of atoms at normal incident to a transmission grating consisting of a standing wave. For atoms with a thermal velocity 10^5 cm/s and a glancing angle 10^{-3} rad, a diffracted by standing evanescent wave beams may be separated by angles of order $10^{-2} - 10^{-3}$ rad.

Another consideration of diffraction of atoms by evanescent wave is a photon picture [Baldwin et al., 1990], in which a diffraction is considered as a result of absorption and emission of photons, leading to a change of atomic momentum. The exponential profile of evanescent wave in direction perpendicular to the interface has contribution from waves of all direction. This permits an atom to acquire, through absorption-stimulated emission process of photons of these waves, a momentum in direction perpendicular to interface for a *specular reflection*.

An atom, interring in evanescent wave, can absorb also a photon from either of two counter propagating waves along an interface vacuum-dielectric. The atom can then reemit that photon by stimulated process back into the same wave. In this case there is zero net change of an atomic momentum parallel to interface. The atom can also reemit photon in to opposite evanescent wave. In this case the atom momentum parallel to the interface will be changed by $2\hbar k$. The absorption and emission of photon pairs changes the momentum in the direction of standing wave but not kinetic energy of atoms due to energy conservation. It means that the modules of a total atomic momentum remains unchanged but the momentum normal to interface must be changed from that in specular reflection. This is a photon explanation of appearance of first *diffraction* order in reflection of atoms by standing evanescent wave. The change of parallel to interface component of atomic momentum on the value $n * 2\hbar k$ gives n'th order of atomic diffraction.

The atom can also leave the evanescent wave in an excited state. In this case the atom acquires from the field an additional energy which equals to the difference between an atomic transition frequency and a photon energy. This kinetic energy changes the atomic momentum only in the direction perpendicular to the surface. In the direction parallel to the surface the atomic momentum can only be changed by one unit of the photon momentum and hence the atom cannot absorb the additional kinetic energy in this direction.

This additional change of energy because of influence of the refractive medium creating of the evanescent wave. This process naturally violates the specular reflection of atoms and also changes the diffraction angles in reflection of atom from standing evanescent wave. The reflection of atoms in a grating configuration of evanescent wave was observed in [Baldwin et al., 1990] but the diffraction pattern was not resolved.

7. ATOM CAVITY

In previous Sections we have described the motion of atoms in a perfectly classical way. We pointed out the necessary conditions (Section 2) for such description. The developed methods of laser cooling and trapping of atoms permits to reach experimental conditions when essential became the quantum states of spatial motion: a quantization of the atomic center of motion or an external coordinate of atom. The quantization of atomic motion in a standing wave [Jessen et al., 1992; Verkerk et al., 1992] and experiment with coherent population trapping [Aspect et al., 1988] are the first examples of such motion. An essential point of such motion is interference of de Broglie waves and occupation with high probability of allowed states of space. In this Section we are going to consider the example of such motion of atom in *atom cavity* where the above pointed properties are the most essential.

The developed technique of laser cooling and trapping of atoms permits nowdays to store atoms for a long time and with high densities [see special issue of scientific journals: Meystre and Stenholm, eds., 1985; Chu and Wieman eds., 1991; Mlynek, Balykin and Meystre, eds., 1992]. One interest of studying the trapping of atoms is the possibility of observing quantum statistical effects, for example the Bose-Einstein condensation of atoms. The Bose-Einstein condensation of spin polarized hydrogen atoms was predicted at high densities and low temperatures [Greytak and Kleppner, 1984]. It was also a several publication in which were analyzed a possibility of Bose-Einstein condensation another atom by applying laser cooling technique [Vigue, 1986; Bagnato et al., 1987]. An alternative approach to the observation of collective quantum effects is a population of *high excited quantum states* by an average more than one atom per state. A laser resonator is well known example of such system: it is quite easy now to reach a population of laser cavity mode by more than one photon.

In this Section will be considered a possibility of creating an atomic cavity which would prove discrete level and the possibility of population of some of these levels by more than one atom per level. Such cavity can be based on the reflection of atoms from mirrors formed by evanescent laser wave (see the previous Section). Fig. 43 shows schematic diagrams of possible atomic cavity configuration, (Balykin and Letokhov, 1989e). These cavities are similar to optical cavities with their material mirrors replaced by light-induced mirrors and instead of photons the cavity is filled by atoms bouncing between these mirrors.

The atomic cavity can be formed even from one mirror: the gravity plays the role of the second mirror and bends the atomic trajectory so that an atom always bounce on the first mirror. In two mirror cavity the gravity plays a little role in the motion of atoms and the calculation of mode structure can be done by using a theory developed for optical resonator. In one mirror cavity (a gravitational cavity) a gravity plays an essential role and a direct application of the formalism developed for optical resonator is difficult. In [Wallis et al., 1992] was given both a classical quantum mechanical calculation the atomic dynamics in the one mirror cavity.

7.1 TWO-MIRROR CAVITY

First let us consider the main properties of such cavity: the maximum and the minimum atomic velocity in the cavity; the cavity stability; the schemes of injection of atoms in a cavity; the maximum atomic density in the cavity.

Atomic velocity in the cavity. The atomic mirrors can effectively reflect atoms whose maximum velocity component normal to the mirror surface $v_\perp < v_{max} = [2U(0)/M]^{1/2}$, where $U(0)$ is the potential energy of the atom at the dielectric-vacuum interface. From this condition the maximum velocity is given by

$$V_{max} \cong \left\{ (v_r v_\gamma)(\gamma/\Omega)q(qP/\pi I_s w_m^2) \right\}^{1/2} \qquad (7.1)$$

where $v_r = \hbar k/M$ is the atomic recoil velocity $v_\gamma = \gamma/k$, and M is the atomic mass. The size of the atomic mirror w_m must be greater than that of the principal atomic field mode on the atomic mirror so as to make the diffraction loss for de Broglie waves insignificant. The waist radius of the principal mode is

$$\sigma_0 = (\lambda_{B\rho} l/4\pi) = [(\lambda l/4\pi)(v_r/v)] \qquad (7.2)$$

where l is the atomic cavity length.

Using the expressions (7.1) and (7.2), one can determine the maximum atomic velocity in the atomic cavity. It depends on the power of laser beam forming the evanescent wave, the detuning frequency of the laser field and the atomic parameters. For instance, for Na atom, a laser power $P = 1\,W$, an atomic mirror size of $w_m = 10\sigma_0$, and a laser field detuning frequency of $\Omega = 10^4 \gamma$, the maximum atomic velocity will be $4.5 * 10^3$ cm/s.

Cavity stability. The principal difference between the optics of material particles and photon optics is that the motion of the particles depends appreciably on the force fields, particularly the earth gravitational field. A simple way to take account of the effect of the gravitational field on the atomic motion in the atomic cavity is to introduce the effective reflective index for de Broglie waves. In the case of vertical cavity the effective reflection index is given by

$$n_{eff}(z) = [1 - V(z)/E] \qquad (7.3)$$

where $V(z)$ is the atomic potential energy in the gravitational field and E is a total atomic energy. The stability condition for a semiconfocal optical cavity is defined by inequality

$$0 \leq (1 - 1/R) \leq 1 \qquad (7.4)$$

where R is the radius of curvature of cavity mirror. The inclusion of the variable refractive index reduces to substitution of the reduced length $l = \int dz/n_{eff}(z)$ for the cavity length in (7.4). In the case of atomic cavity, the reduced length is expressed as

$$L = v_0^2/g - (v_0^2/g)(l - 2g/v_0^2)^{1/2} \qquad (7.5)$$

where v_0 is the atomic velocity at the bottom mirror and g is acceleration due to gravity. The stability condition for the atomic cavity will then take the form

$$0 \leq \left[1 - v_0^2/gR + \left(v_0^2/gR \right) \left(1 - 2gl/v_0^2 \right)^{1/2} \right] \leq 1 \qquad (7.6)$$

The condition (7.6) imposes restrictions on the minimum atomic velocity v_{min} inside the cavity. When $v < v_{min}$, the gravitational field prevents the atoms from reaching the top of the cavity mirror. For example, with the cavity length $l = 5$ cm, v_{min} is 10^2 cm/s. One can also determine the effect of the cavity misadjustment. For the semiconfocal cavity with atomic mirror size of $w_m = 10\sigma_0$ and the atomic velocity of 10^2 cm/s, the angular displacement must not exceed $5 * 10^{-5}$ rad.

The injection. If the atoms are supplied by the thermal source, the width of the atomic velocity distribution is $\delta v \cong v$, where v is an average velocity of atoms in a beam. Both the quantities v and δv are substantially greater than v_{max} and therefore, to reach the necessary injection rate, the atomic beam needs to be preliminary laser retarded and cooled. To increase the proportion of the source atoms undergoing the injection, the atomic beam can be collimated, and to reduce the beam cross section to the waist size of the cavity mode, the beam can be focused (see Section 4) or compressed [Nellessen et al., 1990] by means of radiation pressure force. The estimations show that the preliminary laser cooling and collimation of a thermal atomic beam with an initial divergence of the order of a few degrees can help to give the atomic injection intensities as high as 10^8 atoms/s.

7.2 ONE-MIRROR CAVITY

Atomic velocity in the cavity. In two-mirror cavity the stability requirement demands that the atom in such cavity have a velocity of the order 10^2 cm/s. In one dimensional cavity there is no restriction on the minimum velocity of atoms from point of view stability of their motion and the velocity of atoms can be essentially lower (≈ 10 cm/s). This means that a required laser power for evanescent wave in this case is also much smaller, but an injection in the cavity atomic beam should be also cooled to a smaller temperature.

Cavity stability. To study stability of mirror cavity one can also use an approach analogous to the study of stability of optical resonator. The stable motion of atom is possible if the apex of atomic trajectory

$$h < R/2 \tag{7.7}$$

where R is the the radius of curvature of the mirror, $h = v_z^2/2g$ and v_z is the vertical velocity component on the mirror.

Injection. In order to fill one mirror cavity was proposed [Wallis et al., 1992] to use an atomic cloud produced by a optical molasses or in a magnetooptic trap. In such scheme of injection at least some of the space and velocity variables of atomic cloud should be inside a stability region of the trap. The ideal situation is one in which the initial atomic phase-space is a Gaussian wave packet both in real and velocity space. In this case the position and momentum satisfy the Heisenberg uncertainty relation $\delta r * \delta p = \hbar/2$. The population of transverse ground mode P_{tr} of one mirror cavity filled from atomic cloud:

$$P_{tr} = \hbar^2 / \left[\left(\delta r^2 + \sigma_0^2/2 \right) \left(\delta p^2 + \hbar^2/2\sigma_0^2 \right) \right] \tag{7.8}$$

where σ_0 is the waist the transverse atomic mode. The population of transverse ground mode can be equal to $P_{tr} = 1$ if the initial Gaussian distribution ($\delta p*\delta r = \hbar/2$) matches to the transverse mode ($r = \sigma_0/2^{1/2}$).

In real experimental situation an atomic cloud of Cs [Foot and Sreane, 1990] obtained in magnetooptic trap has a radius $\delta r \approx 50$ μm and a velocity spread 2 cm/s. This corresponds to $\delta r * \delta p = 2000$ $\hbar$, which is considerably larger then for Gaussian wave packet. The loading the one mirror trap from such magnetooptic trap will permit to get the population of transverse ground state only $P_{tr} = 2.5 * 10^{-7}$. Fig. 44 shows the calculated density profile for the eigenmode corresponding to the 16th excited state of the longitudinal motion of atom in one mirror cavity. a): Ground state of transverse motion; (b): one of the excited transverse state of transverse motion [Wallis et al., 1992].

7.3 LIFETIME OF ATOMS IN THE CAVITY

The lifetime of atoms in the cavity is one of the most important parameter of the cavity because it determines the maximum density of atoms in the cavity and the

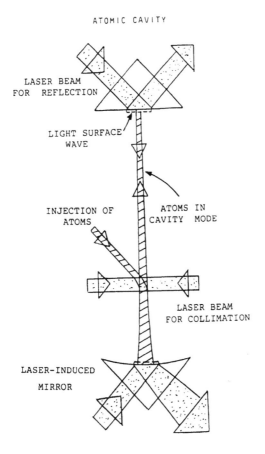

Fig. 43 Schematic diagram of atomic cavity with light induced mirrors. Cavity is based on the reflection of atoms from mirrors formed by evanescent laser wave [Balykin and Letokhov, 1989e].

value of degeneracy factor. The atomic lifetime in the cavity depends mainly on the following processes: (a) momentum diffusion, (b) loss of atoms due to the diffraction by the atomic mirrors, (c) tunneling of atoms to the dielectric surface, followed their absorption or diffusive reflections and (d) scattering of atoms by the residual gas particles and the atoms themselves in the cavity.

Momentum diffusion. When an atom moves in the evanescent light wave, it may reemit a photon and thus change its momentum. The angular indeterminacy of the atomic trajectory due to this reemission of the photon is $\delta\phi_\perp \cong v_r/v$, where v is the atomic velocity. The angular divergence of atomic wave field in the case of single transverse cavity mode localization is $\delta\phi_M = w/r$, where w is the waist radius of the principal mode of the cavity, r is the radius of curvature of the de Broglie wave on the atomic mirror. After reemission of a photon an atom will remain in the mode only if

$$\delta\phi_\perp/\delta\phi_M \le 1 \qquad (7.9)$$

This condition leads to the following expression for the momentum-diffusion limited atomic lifetime:

$$t_{difs} = 8\pi \tau_{sp}(\Omega^2/\omega_{R0}^2)(1/x_0) \qquad (7.10)$$

where ω_{R0} the Rabi frequency on the dielectric – vacuum interface, x_0 – the characteristic length of evanescent wave (see Section 6). For the above used parameters of atomic cavity the momentum-diffusion-limited lifetime is about 50 sec.

Diffraction losses. To estimate diffraction losses of atoms, use can be made of analogy between the atomic cavity and optical cavity, the light wavelength in the expression for the loss coefficient being replaced by the de Broglie wavelength. Then the characteristic atomic lifetime in the cavity is

$$t_{dif} = 1/v\alpha_{dif} \qquad (7.11)$$

where α_{difr} the diffraction loss coefficient for atoms in the cavity. For an atomic confocal cavity with an atomic mirror size of $w_m = 10\sigma_0$, the diffraction losses are negligible.

Tunneling. When an atom moves with a velocity close to the maximum velocity v_{max} (see (7.1)), it may undergo tunneling over the potential barrier of the surface light wave. In this case atom is lost from the cavity. The tunneling limited atomic lifetime in the cavity mode is

$$t_{tun} = 1/v\Theta, \qquad (7.12)$$

where Θ is the transmisitivity of the potential barrier of the surface light wave. For the case $v \le v_{max}$, the transmitivity is [Kemble, 1935]:

$$\Theta = \left\{ 1 + \exp\left[2\int_{z_1}^{z_2} Q(z)dz \right] \right\}^{-1} \qquad (7.13)$$

where $Q^2(z) = 2M[U(z) - E]/\hbar^2$., and z_1, z_2 are the turning points. With a characteristic penetration depth of the surface light $\lambda/2$ and $v/v_{max} = 0.8$ the tunneling lifetime is $t_{tun} \cong 8 * 10^3$ s.

Collisional scattering of atoms. The main limitation on the maximum atomic lifetime is imposed by mutual scattering of the localized atoms. The steady state concentration N_{at} in the cavity is related to the intensity J of the atomic beam being injected into the cavity and to the atomic lifetime through the collisions in the cavity by the simple relation

$$N_{at} = (J/Sv)(t_{at}v/l) = Jt_{at}/V, \qquad (7.14)$$

where S and V are the cross-sectional area and the volume of the atomic beam in the cavity, respectively. According to above estimates, t_{at} is determined mainly by momentum diffusion lifetime. But the atomic concentration N_{at} continues to grow in proportion to the atomic beam intensity J only until atomic collisions inside the cavity become important, or in other words, the maximum atomic concentration is reached when the scattering lifetime becomes equal to the momentum — diffusion atomic lifetime:

$$1/t_{scat} = N_{max}\sigma_v v \approx 1/t_{difs} \qquad (7.15)$$

where σ_v is scattering cross-section of the atoms moving with velocity v. The estimated (Balykin and Letokhov, 1989e) maximum atomic concentration and the injected atomic beam intensity to achieve such concentration in the cavity are $N_{max} = 1.1 * 10^9$ cm^{-3}, $J = 2.9 * 10^2$ atoms/s.

The atomic beam density in the cavity could be substantially increased by using the circular cavity configuration: the role of atomic collisions in this cavity becomes important at more higher atomic concentrations due to the unidirectional atomic motion. In such cavity the maximum concentration could be $N = 2.2 * 10^{11}$ atoms/s. Another solution for avoiding collision is to insert mode selector into the cavity which will remove the atoms from most of the modes except the basic one. This permit to increase the collision live time of atoms and increase the degeneracy parameter.

7.4 DEGENERACY FACTOR OF THE CAVITY

The atomic cavity considered here is largely analogous to a laser cavity with a high concentration of photons in the mode. One of the main parameters characterizing the field in laser cavity is its degeneracy. Let us estimate the atomic degeneracy attainable in the atom cavity. The degeneracy in the atom cavity plays also a role analogous to the one defined for Bose-Einstein condensation. The atoms localized in the principal

+F   +F

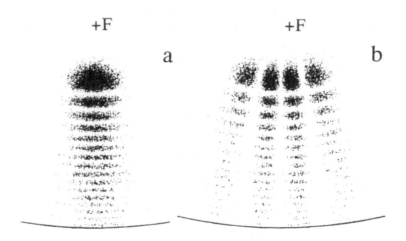

Fig. 44 Density profile for the eigenmode corresponding to the 16th excited state of the longitudinal motion of atom in one mirror atomic cavity. (a): Ground state of transverse motion; (b): one of the excited transverse state of transverse motion. F is the focus of the mirror [Wallis et al., 1992].

cavity mode can be considered to be confined in one-dimensional potential well. The energy of the nth level of such well is

$$E_n = (\pi^2 \hbar^2 / 8Ml^2)n^2 \qquad (7.16)$$

The number of levels in the energy interval δE, hence with velocity δv is

$$\delta n = (4Ml^2 / \pi^2 \hbar^2 n)\delta E = (4Ml/\hbar)\delta v \qquad (7.17)$$

We determine the atomic degeneracy as the average number of atoms on the certain allowed sublevel:

$$g_{at} = n_{at}/\delta n \qquad (7.18)$$

where n_{at} is the total number of atoms in the cavity. The cavity degeneracy in the terms of atomic concentration in the cavity will be

$$g_{at} = N(\hbar^2 / 16M^2)(l/v\delta v) \qquad (7.19)$$

The critical atomic concentration in the cavity at which the degeneracy can be larger than > 1 is

$$N_{cr} \cong (\delta vv/v_{rec}^2)\lambda^2 l \qquad (7.20)$$

For the typical atomic cavity parameters, considered above, $v = 2 * 10^2$ cm/s, $\delta v = 1$ cm/s, $v_r = 3$ cm/s, $\lambda = 6 * 10^{-5}$ cm and $l = 5$ cm, the critical atomic concentration $N_{cr} = 1.2 * 10^9$ at/cm^3. At such concentration the degeneracy can reach unity for linear cavity and could be much higher for the ring cavity. It is interesting to compare the critical density in atomic cavity and in an isotropic case. The critical atomic concentration in an isotropic source ($\delta v = v$, $\delta\phi = 4\pi$ sr and at $\lambda_{Br} = 1$ Å) is $N_{cr} = 1/\lambda_{Br}^3 = 10^{24}$ at/cm^3, which is many orders of magnitude greater than the atomic concentration in the cavity. The degeneracy $g_{at} > 1$ at low atomic concentration in the cavity is achieved on account of the velocity monochromatization and their accumulation within narrow solid angle of the cavity mode.

7.5 EXPERIMENT WITH GRAVITATIONAL CAVITY

From an experimental point of view one of the simplest configurations for a cavity for atoms is the gravitational cavity [Aminoff et al., 1993]. It consists of an evanescent wave on horizontally aligned parabolic glass surface at its lower end to reflect the falling atoms upward and using gravity to turn the atoms back at its upper end. The curved reflector is refocused the atomic trajectories. A cloud of cesium atoms, trapped and laser cooled in a magnetic trap, was released a few millimeters above the surface and falls on the reflector. Atoms bouncing on the evanescent wave are subsequently detected by fluorescence. Advantage of using an atom with larger mass, such, as cesium, is that the recoil velocity is smaller. This gives a lower recoil limited temperature, leading to a slower thermal expansion of the atomic cloud. In experiment 10^7 atoms were loaded in a cloud with a diameter of about 1 mm. The atomic reflector was a parabolic surface with radius of curvature 20 mm, which was placed about 3 mm below the center of the trap.

Figure 45 shows the observations of several successive reflection of Cs atoms from the atom parabolic mirror. Curve (a) is the fluorescent signal from Cs atoms present around the release point as a function of the delay after release from trapped cloud. Delay varied from 0 to 200 ms. (b) Measurement extended from 120 ms to 220 ms. The curves show four successive bounces separated by the round trip time 50 ms. The lifetime of the cavity and the number of bounces was limited by the loss of atoms due to optical pumping and heating through absorption and emission.

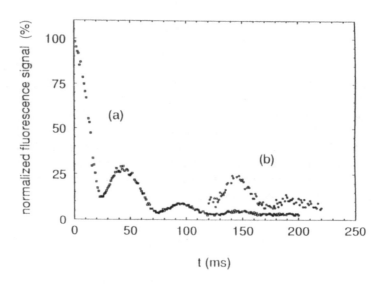

Fig. 45 Figure shows the observations of several successive reflection of Cs atoms from the atomic parabolic mirror. Curve (a) is the fluorescent signal from Cs atoms present around the release point as a function of the delay after release from trapped cloud. Delay varied from 0 to 200 ms. (b) Measurement extended from 120 ms to 220 ms. The curves show four successive bounces separated by the round trip time 50 ms (Aminoff et al., 1993).

8   CONCLUSION

Nowadays by using the mechanical forces exerted on atoms by light it is possible to achieve both cooling and trapping of atoms. Presently more than twenty laboratories in the world are working in this field. The minimum temperature of atoms in laser cooling experiment and maximum density of atoms in the atomic traps permit to orient the latest research works towards such ambitious goals as a realization of Bose-Einstein condensation of cold atoms, atomic interferometry, improvement of atomic clock. Another interesting possibility of using light pressure forces is the development of a new kind of optics — the optics of neutral atomic beams. In this review we have discussed a number of recent experiments which have demonstrated the possibility to develop several elements of this optics: *atom optics with laser light.*

REFERENCES

Aminoff, C.G., Bouyer, P. and Desbiolles, P., (1993), C.R. Acad. Sci., Paris, t.316, Seria II, 1535.

Anderson, A., Haroche, S., Hinds, E.A., Jhe, W., Meschede, D. and Moi L., (1986), Phys. Rev. **A34**, 3513.

Arimondo, E., (1992), Int. School of Physics "Enrico Fermi", ed. Arimondo, E., Phillips W. and Strumia F., *Laser Manipulation of Atoms and Ions*, 191.

Ashkin, A., (1970), Phys. Rev. Lett. **25**, 1321.

Ashkin, A., (1978), Phys. Rev. Lett. **40**, 729.

Ashkin, A., (1980), Science, **210**, 1081.

Askarian, G.A., (1962), Sov. Phys. JETP **15**, 1088.

Askarian, G.A., (1973), Usp. Fiz. Nauk., **110**, 115 (in Russian).

Aspect, A., Dalibard, J., Heidmann, A., Salomon, S. and Cohen-Tannoudji, C., (1986), Phys. Rev. Lett. **57**, 1688.

Aspect, A., Arimondo, E., Kaiser, R., Vansteenkiste, N., Cohen-Tannoudji, C., (1988), Phys. Rev. Lett. **61**, 826.

Aspect, A., Arimondo, E., Kaiser, R., Vansteenkiste, N., Cohen-Tannoudji, C., (1989), J. Opt. Soc. Am. **B6**, 2112.

Aspect, A., Vansteenkiste, N., Kaiser, R., Haberland, H. and Karrais, M., (1990), Chem. Phys., **145**, 307.

Bagnato, V., Pritchard, D.E. and Kleppner, D., (1987), Phys. Rev. **A35**, 4354.

Baklanov, E.V. and Dubezkii, B.J. (1976), Opt. Spectr., **41**, 3.

Baldwin, K.G.H., Hajnal, J.V., Fisk, P.T.H., Bachor, H.-A. and Opat, G.I.J., (1990), J. Mod. Opt., **37**, 1839.

Balykin, V.I., Sidorov, A.I., Letokhov, V.S., (1984), Pis'ma Zh. Eksp. Teor. Fiz. **40**, 251.

Balykin, V.I., Letokhov, V.S., Minogin, V.G., Rozhdestvensky, Yu.V., Sidorov, A.I., (1985), J. Opt. Soc. Am., **B2**, 1776.

Balykin, V.I., Letokhov, V.S., Sidorov, A.I., (1986) Pis'ma Zh. Eksp. Teor. Fiz. **43**, 172.

Balykin, V.I., Sidorov V.I., (1987a), Appl. Phys. **B42**, 51.

Balykin, V.I., Letokhov, V.S., (1987b), Opt. Commun. **64**, 151.

Balykin, V.I., Letokhov, V.S., Sidorov, A.I., Ovchinnikov, Yu.B., (1987d), Pis'ma Zh. Eksp. Teor. Fiz. **45**, 282 (Sov. Phys. JETP Lett., **45**, 353).

Balykin, V.I., Letokhov, V.S., Sidorov, A.I., Ovchinnikov, Yu.B., (1988a), J. Mod. Opt. **35**, 17.

Balykin, V.I., Letokhov, V.S., Ovchinnikov, Yu.B., Sidorov A.I. and Shulga S.V., (1988b), Opt. Lett., **13**, 958.

Balykin, V.I., Letokhov, V.S., Ovchinnikov, Yu.B., Sidorov, A.I., (1988c), Phys. Rev. Lett. **60**, 2137 (Errata **61**, 902).

Balykin, V.I., Letokhov, V.S., (1988d) Sov. Phys. JETF **67**, 78.

Balykin, V.I., Lozovik, Yu.E., Letokhov V.S., Ovchinnikov, Yu.B., Sidorov, A.I., Shulga, S.V. and Letokhov, V.S. (1989a), JOSA, **B6**, 2178.

Balykin, V.I., Zueva, T.V., Sidorov, A.I., (1989b) Sov. J. Quantum Elect. **19**, 1502.

Balykin, V.I., Lozovik, Yu.E., Ovchinnikov, Yu.B., Sidorov A.I., Shulga, S.V.,and Letokhov, V.S., (1989c), J. Opt. Soc. Am., **B6**, 2178.

Balykin, V.I., (1989d), Appl. Phys. **B49**, 383.

Balykin, V.I., Letokhov, V.S., (1989e), Appl. Phys. **B48**, 517.

Balykin, V.I., Letokhov, V.S., (1989f), Phys. Today, **4**, 23.

Balykin, V.I., (1989g), Appl. Phys. **B49**, 383.

Balykin, V.I., Letokhov, V.S., Ovchinnikov, Yu.B., Sidorov, A.I., (1989h), Pis'ma Zh. Eksp. Teor. Fiz., **49**, 383.

Balykin, V.I., Letokhov, V.S., Ovchinnikov, Yu. B. and Shulga, S.V., (1990), Opt. Commin., **77**, 152.

Benewitz, G. and Paul W., (1954), Zs. f. Phys., **139**, 489.

Benewitz, G., Paul, W. and Schlier C., (1955), Zs. f. Phys., **141**, 6.

Bernstein, R.B., (1982), *Chemical Dynamics via Molecular Beam and Laser Techniques*. (Cladendon Press: Oxford).

Berkhout, J.J., Luiten O.J., Setija, I.D., Hijmans, T.W., Mizusaki, T., Walraven, J.T.M., (1989), Phys. Rev. Lett., **63**, 1689.

Bigelow, N.P. and Prentiss, M.G., (1990), Phys. Rev. Lett., **65**, 555.

Bjorkholm, J.E., Freeman, R.E., Ashkin, A.A., Pearson, D.B., (1978), Phys. Rev. Lett. **41**, 1361.

Bjorkholm, J.E., Freeman, R.E., Ashkin, A.A., Pearson, D.B., (1980), Opt. Lett., **5**, 111.

Bonch-Bruyevich, A.M., Khodovoi, A.A., (1968), Sov. Phys. Usp., **10**, 637.

Bonze, U., Rauch, H., eds., (1979), *Neutron Interferrometry*. (Cladendon Press: Oxford).

Born, M., Wolf, E., (1984), *Principles of Optics*, (Pergamon Press: Oxford).

Carnal, O., Faulstich, A. and Mlynek, J., (1991), Appl. Phys., **B53**, 88.

Carnal, O. and Mlynek, J., (1991), Phys. Rev. Lett. **66**, 2689.

Carnal, O., Sigel, M., Sleator, T., Takuma, H. and Mlynek, J., (1991), Phys. Rev. Lett. **67**, 3231.

Carver, T.R., (1961), Phys. Rev., **124**, 800.

Chen, J., Story, J.G., Tollett, J.J. and Hulet, R.G., (1992), Phys. Rev. Lett., **69**, 1344.

Chen, J., Tollett, J.J. and Hulet, R.G., (1993), Phys. Rev. **A47**, 2128.

Chu, S., (1991) Science, **253**, 861.

Chu, S. and Wieman, C., eds. (1989), J. Opt. Soc. Am., **B6**, Special Issue No. B11, "*Laser Cooling and Trapping of Atoms*".

Chu, S., Weiss, D.S., Shevy, Y. and Ungar, P., (1991), in *Atomic Physics II*, Haroche, S., Gay, J.C. and Grynberg, G., eds., (World Scientific Publishing Co., Singapore) p. 633.

Cohen-Tannoudji, C., (1951), C.R. Acad. Sci. (Paris), **252**, 394.

Cohen-Tannoudji, C. and Reymond, S., (1978), Eberly J. and Lambropoulos P., eds., in *Multiphoton Processes* (N.Y. Wiley), p. 103.

Cohen-Tannoudji, C. and Phillips, W.D., (1990), Physics Today, **10**, 33.

Cohen-Tannoudji, C., (1991), in *Fundamental Systems in Quantum Optics*, (Elsevier Science Publ., B.V.).

Cook, R.J., (1980), Phys. Rev. Lett., **44**, 976.

Cook, R.J., Hill, R.K., (1982), Opt. Commin., **43**, 258.

Dalibard, J. and Cohen-Tannoudji, C., (1985), J. Opt. Soc. Am., **B2**, 1707.

Dalibard, J. and Cohen-Tannoudji, C., (1989), J. Opt. Soc. Am., **B6**, 2023.

Dalibard, J., Heidman, A., Salamon, S., Aspect, A., Metcalf, H., Cohen-Tannoudji, C.: In *Fundamental of Quantum Optics* II, F. Ehlotzky ed., (Springer Series Lectures Notes in Physics).

Dalibard, J., Salamon, S., Aspect, A., Metcalf, H., Heidman, A. and Cohen-Tannoudji, C., (1987), in *Laser Spectroscopy VIII*, eds. Svanberg, S. and Pearson, W. (Springer Verlag, Heidelberg) p. 81.

Dalibard, J., Salomon, S., Aspect A., Arimondo, E., Kaiser, R., Vansteenkiste, N. and Cohen-Tannoudji, C., (1991), in *Atomic Physics II*, Haroche, S., Gay, J.C. and Grynberg, G., eds., (World Scientific Publishing Co., Singapore), p. 192.

Dehmelt, H., (1976), Nature, **262**, 777.

Dehmelt, H., (1983), in *Advances in Laser Spectroscopy*, Arecchi, F.T., Strumia, F. and Walther, H., eds., (Plenum Press, New York) p. 153.

Dicke, R., (1953), Phys. Rev., **89**, 472.

Einstein, A., (1909), Phys. Zs., **10**, 185; ibid. (1917) **18**, 121.

Estermann, L. and Stern, O., (1930), Zs. f. Phys., **61**, 95.

Esslinger, T., Weidenmüller, M., Hemmerich, A. and Hänsch, T., (1993), Opt. Lett., **18**, 450.

Feynmann, R.F. and Hibbs, A.R., (1965), *Quantum Mechanics and Path Integrals* (McGraw-Hill: N.Y.).

Feron, S., Reihardt, J., Le Boiteux, S., Gorceix, O., Baudon, J., Ducloy, M., Robert, J., Miniatura, Ch., Nic Chormaic, S., Haberland, H. and Lorent, V., (1993), Opt. Commin., **102**, 83.

Frank, A.I., (1987), Usp. Fiz. Nauk., **151**, 229; (Sov. Phys. Uspekhi, **30**, 110).
Frank, A.I., (1991), Usp. Fiz. Nauk., **161**, 95; (Sov. Phys. Uspekhi, **34**, 980).
Friedburg, H., und Paul, W., Naturwiss., (1950), **37**, 20.
Friedburg, H., und Paul, W., Naturwiss., (1951), **38**, 159.
Friedburg, H., und Paul, W., (1951a), Zs. f. Phys., **130**, 493.
Frish, O.R. and Stern, O., (1933), Zs. f. Phys., **84**, 430.
Frish, O.R., (1933), Zs. Phys., **86**, 42.
Foot, C. and Steane, A., (1990), Europhys. Lett., **14**, 231.
Gallatin, G., Gould, P.J., (1991), Opt. Soc. Am., **B8**, 502.
Goldenberg, H.M., Kleppner, D. and Ramsey, N.F., (1960), Phys. Rev. Lett., **8**, 361.
Goldenberg, H.M., Kleppner, D. and Ramsey, N.F., (1962), Phys. Rev. Lett., **126**, 603.
Gordon, J.P., Zeiger, H.J. and Townes, C., (1954), Phys. Rev., **95**, 282.
Gordon, J.P., Zeiger, H.J. and Townes, C., (1955), Phys. Rev., **99**, 1264.
Gordon, J.P. and Ashkin, A., (1980), Phys. Rev., **A21**, 1606.
Gould, P.I., Ruff, G.A. and Pritchard, D.E., (1986), Phys. Rev. Lett., **56**, 827.
Greytak, T.J. and Kleppner, D., in Les Houches, Session XXXVIII, 1982 – *New Trends in Atomic Physics*, G. Grynberg and R. Stora, eds. (North Holland, Amsterdam, 1984) p. 1125.
Grimm, R., Ovchinnikov, Yu.B., Sidorov, A.I. and Letokhov, V.S., (1990), Phys. Rev. Lett., **65**, 1415.
Grimm, R., Letokhov, V.S., Ovchinnikov, Yu.B., and Sidorov, A.I., (1991), Pis'ma Zh. Eksp. Teor. Fiz., **54**, 611 (JETP Lett., **54**, 615).
Grimm, R., Ovchinnikov, Yu.B., Sidorov, A.I. and Letokhov, V.S., (1991), Opt. Commun., **84**, 18.
Grivet, P., (1972), *Electron Optics*, sec. edit. (Oxford Univ. Press).
Gould, P.I., Ruff, G.A. and Pritchard, D.E., (1986), Phys. Rev. Lett., **56**, 827.
Hajnal, J.V. and Opat, G.I., (1989), Opt. Commin., **71**, 119.
Hänsch, H.J. and Schawlow, A.L., (1975), Opt. Comm., **13**, 68.
Hemmerich, A., Schropp, D.Jr. and Hänsch, T.W., (1991), Phys. Rev., **44**, 1910.
Hemmerich, A., Zimmerman, C. and Hänsch, T.W., (1993), Europhys. Lett., **22**, 89.
Javanainen, J., (1990), Phys. Rev. Lett., **64**, 519.
Jessen, P.S., Gerz, C., Lett, P.D., Phillips, W.D., Rolston, S.L., Spreeuw, R.J.C. and Westbrook, C.I., (1992), Phys. Rev., **69**, 49.
Kasevich, M., Weiss, D. and Chu, S., (1990), Opt. Lett., **15**, 607.
Kazantsev, A.P., (1974), Zh. Eksp. Teor. Fiz., **66**, 1599 (Sov. Phys. JETP **39**, 784).
Kazantsev, A.P., Ryabenko, G.A., Surdutovich, G.I. and Yakovlev, V.P., (1985), Phys. Rep., **129**, 75.
Kazantsev, A.P., Smirnov, V.S., Surdutovich, G.I., Chudesnikov, D.O. and Yakovlev, V.P. (1985), J. Opt. Soc. Am., **B2**, 1731.
Kazantsev, A.P., Chudesnikov, D.O. and Yakovlev, V.P., (1986), Sov. Phys. JETF **63**, 951.
Kazantsev, A.P. and Krasnov, I.V., (1987), Pis'ma Zh. Eksp. Teor. Fiz., **46**, 333 (JETP Lett., **46**, 420).
Kazantsev, A.P. and Krasnov, I.V., (1989), J. Opt. Soc. Am., **B6**, 2140.
Kazantsev, A.P., Surdutovich, G.I. and Yakovlev, V.P., (1991), *Mechanical Action of Light on Atom*, (World Scientific Publishing Co., Singapore).
Kearney, P.D., Klein, A.G., Opat, G.I., Gähler, R., (1980), Nature, **287**, 313.
Keith, D.W., Schattenburg, M.L., Smith, H.I. and Pritchard, D.E., (1988), Phys. Rev. Lett., **61**, 1580.
Keith, D.W., Ekstrom, C.R., Turchette, Q.A. and Pritchard, D.E., (1991), Phys. Rev. Lett., **66**, 2693.
Kemble, E.C., (1935), Phys. Rev., **48**, 549.
Knauer, F. and Stern, O., (1929), Zs. f. Phys., **53**, 799.
Korsynskii, M.I. and Fogel, Ya.M., (1951), Zh. Eksp. Teor. Fiz., **21**, 25; ibid. (1951), **21**, 38.
Landay, L.D., Livshits, E.M. (1985). *Quantum Mechanics* (Addison-Wesley, Reading Mass.).
Letokhov, V.S., (1968), Pis'ma Zh. Eksp. Teor. Fiz., **7**, 348 (in Russian); [Sov. Phys. JETP Lett., (1968), **7**, 272].

Letokhov, V.S. and Minogin, V.G. and Pavlik, B., (1977), Zh Eksp. Teor. Fiz., **72**, 1318 (Sov. Phys. JETP **45**, 698).

Letokhov, V.S. and Minogin, V.G., (1978), Zh. Eksp. Teor. Fiz., **74**, 1318 (Sov. Phys. JETP **47**, 690).

Letokhov, V.S. and Minogin, V.G., (1978), Appl. Phys., **17**, 99.

Letokhov, V.S. and Minogin, V.G., (1981), Physics Reports, **73**, 1.

Lett, P., Watts, R., Westbrook, C., Phillips, W.D., Could, P., Metcalf, H., (1988), Phys. Rev. Lett., **61**, 169.

Martin, P.J., Oldaker, B.G., Miklich, A.H. and Pritchard, D.E., (1988), Phys. Rev. Lett., **60**, 516.

Mauri, F. and Arimondo, E., (1991), Euro. Phys. Lett., **16**, 717.

Mauri, F., Popoff, F. and Arimondo, E., (1991a), in *Light Induced Kinetic Effects* p. 89 (ETS Editrice, Piza), eds. Moi, L., Gozzini, S., Gabbanni, C., Arimondo, E. and Strumia, F.

McClelland, J.J., Scheinfein, M.P., (1991), J. Opt. Soc. Am., **B8**, 1974.

Meystre, P. and Stenholm, S., eds., (1985), J. Opt. Soc. Am., Special Issue No. B2, N. 11, *"Mechanical Effects of Light"*.

Meyer, D.T., Meyer, H., Hallidy, W. and Kellers, C.F., (1963), Cryogenics **3**, 150.

Minogin, V.G., Serimaa, O.T., (1979), Opt. Commun, **30**, 373.

Minogin, V.G., Letokhov, V.S., (1987), *Laser Radiation Pressure on Atoms* (Gordon and Breach, New York).

Minogin, V.G. and Rozhdestvensky, Yu.V., (1987), Zh. Eksp. Teor. Fiz., **93**, 1173.

Mizushima, M., (1964), Phys. Rev., **A133**, 414.

Mlynek, J., Balykin, V.I. and Meystre, P., eds., (1992), J. Opt. Soc. Am., Special Issue **B54**, No 5, *"Optics and Interferometry with Atoms"*.

Moi, L., Gozzini, S., Gabbanini, C., Arimondo, E, and Strumia, F., eds. (1991), *Light Induced Kinetic Effects*, (ETS Editrice, Piza).

Nayak, V.U., Edwards, D.O. and Masuhara, N., (1983), Phys. Rev. Lett., **50**, 990.

Nellesen, J., Müller, J.H., Sengstock, K. and Ertmer, W., (1989), J. Opt. Soc. Am., **B6**, 2149.

Nellessen, J. Werner, J. and Ertmer, W., (1990), Opt. Commin., **78**, 300.

Nienhuis, G., Van der Straten, P., Shang, S-Q., (1993), Phys. Rev., **A44**, 462.

Ol'shani, M.A. and Minogin, V.G. (1991), in "Proceedings LIKE Workshop", edt. L. Moi et al (ETS Editrice, Pisa), p. 99–110.

Ol'shani, M.A., Letokhov, V.S. and Minogin, V.G. (1992), Mol. Cryst. Liq. Cryst. Sci. Technol. – Sec B: Nonlinear Optics, Vol. 3, p. 283.

Opat, G.I., Wark, S.J. and Cimmino, A., (1992), Appl. Phys., **B54**, 396.

Ovchinnikov, Yu.B., Shulga, S.V. and Balykin, V.I. (1991), J. Phys: At. Mol. Opt. Phys., **24**, 3173.

Papoulis, M., (1968), *System and Transforms with Application in Optics* (McGraw, N.Y.).

Phillips, W., Gould, P.I., Lett, P.D., (1988), Science, **239**, 877.

Prentiss, M.G. and Ezekel, S., (1986), Phys. Rev. Lett., **56**, 46.

Prentiss, M., Cable, A. and Bigelow, N.P., (1989), J. Opt. Soc. Am., **6**, 2155.

Pritchard, D.E. *XII International Conference on Atomic Physics*. eds., J. Zorn and R.R. Lewis. (American Institute of Physics, New York, N.Y., 1991), p. 165.

Pritchard, D.E. and Oldaker, B.G., in *Coherence and Quantum Optics VI*, ed., J.H. Eberly, L. Mandel and E. Wolf (Plenum, New York, N.Y., 1990), p. 937.

Pussep, A.Y., (1976), Zh. Eksp. Teor. Fiz., **79**, 851.

Raab, E.L., Prentiss, M., Cable, A., Chu, S. and Pritchard, D., (1987), Phys. Rev. Lett., **59**, 2631.

Raether, H., (1988), *Surface Plasmons*, (Springer, Berlin).

Ramsey, N.F., (1956), *Molecular Beams* (Clarendon, Oxford).

Rigrod, W.W., (1963), Appl. Phys. Lett., **2**, 51.

Riis, E., Weiss, D.E., Moler, A. and Chu, S., (1990), Phys. Rev. Lett., **64**, 1658.

Ruska, E., (1980), *The Early Development of Electron Lenses and Electron Microscopy*. (S. Hirzel Verlag: Stuttgart).

Salomon, S., Dalibard, J., Aspect, A., Metcalf, H., Cohen-Tannoudji, C., (1987), Phys. Rev. Lett., 59, 1659.

Schmahl, G. and Rudolph, D. eds. (1984), *X-Ray Microscopy*, Springer Series in Optical Sciences, 43, (Springer Berlin, Heidelberg, New York).

Sears, V.F., (1989), *Neutron Optics*, New York.

Seifert, W., Adams, C.S., Balykin, V.I., Heine, C., Ovchinnikov, Yu. and Mlynek, J., (1994), Phys. Rev., A49, 3814.

Shang, S.-Q., Sheeby, B., van der Straten, P. and Methcalf, H. (1990), Phys. Rev. Lett., 65, 317.

Shapiro, F.L., (1976), *Neutron Research*. (Publ. House "Nauka", Moskau).

Shevy, Y., Weiss, D.S., Ungar, P.J. and Chu, S., (1989), Phys. Rev. Lett., 62, 1118.

Shevy, Y., Weiss, D.S. and Chu, S., (1989), in *Proceedings of Conference on Spin Polarized Quantum Systems* (World Scientific, Singapore), p. 287.

Sheehy, B., Shang, S.-Q., Watts, R., Hatamian, S. and Metcalf, H., (1989), J. Opt. Soc. Am., 6, 2165.

Sheehy, B., Shang S.-Q., van der Straten, P., Hatamian, S. and Metcalf, H., (1990), Phys. Rev. Lett., 64, 858.

Shimizu, F., Shimizu, K., Takuma, H., (1990), Chem. Phys., 145, 327.

Shutz, G., Steyerl, A., Mampe, W., (1980), Phys. Rev. Lett., 44, 1400.

Sleator, T., Pfau, T., Balykin, V. and Mlynek, J., (1992), Appl. Phys., B54, 375.

Stenholm, S., (1986), Rev. Mod. Phys., 58, 699.

Stern, O., (1929), Naturwiss., 17, 391.

Szilgyi, M., (1988), *Electron and Ion Optics* (Plenum Press: N.Y.).

Timp, G., Behringer, R.E., Tennant, D.M., Cunningham, J.E., Prentiss, M. and Berggren, K.K. (1992), Phys. Rev. Lett., 69, 1636.

Tollett, J.J., Chen, J., Story, J.G., Ritchie, N.W.M., Bradley, C.C. and Hullet, R.G., (1990), Phys Rev. Lett., 65, 559.

Ungar, P.J., Weiss, D.S., Riis, E. and Chu, S., (1989), J. Opt. Soc. Am., B6, 2058.

Valentin, C., Gagné, M.-C., Yu.J. and Pillet, P., (1992), Europhys. Lett., 17, 133.

Vanthier, R., (1949), Comp. Rend., 228, 1113.

Verkerk, P., Lounis, B., Salomon, C. and Cohen-Tannoudji, C., (1992), Phys. Rev. Lett., 68, 3861.

Vigué, J., (1986), Phys. Rev., A34, 4476.

Walraven, J.T.M., (1992), in Les Houches, Session LIII, 1990, *Fundamental Systems in Quantum Optics*, Dalibard, J., Raimond, J.M. and Zinn-Justin, J., eds. (North Holland, Amsterdam).

Wallis, H., Dallibard, J. and Cohen- Tannoudji, C., (1992), Appl. Phys., B54, 407.

Wang, Y., Cheng, Y. and Gai, W., (1989), Opt. Commin., 70, 462.

Wang, Y., Gai, W.Q., Cheng, Y.D., Liu, L., Lio, Y., Luo, Y. and Zhang, X.D., (1990), Phys. Rev., 42, 4032.

Westbrook, C.I., Watts, R.N., Tanner, C.E., Rolston, S.L., Phillips, W.D., Lett, P.D. and Gould, P.L., (1990), Phys. Rev. Lett., 65, 33.

Wineland, D.J. and Dehmelt, T.W., (1975), Bull. Amer. Phys. Soc., 20, 637.

Wineland, D.J. and Itano, W.M., (1979), Phys. Rev., A20, 1521.

Wineland, D.J. and Itano, W.M., (1987), Physics Today, 6, 34.

INDEX

LASER SCIENCE AND TECHNOLOGY
AN INTERNATIONAL HANDBOOK

SECTIONS
Distributed Feedback Lasers
Laser Fusion
Lasers and Surfaces
Laser Photochemistry
Chaos and Laser Instabilities
Excimer Lasers
Lasers and Nuclear Physics
Lasers in Medicine and Biology
Lasers and Fundamental Physics
Laser Spectral Analysis
Lasers and Communication
Solid State Lasers
Laser Diagnostics in Chemistry
Semiconductor Diode Lasers
Topics in Theoretical Quantum Optics
Gas Lasers
Fiber Optics Devices
Ultrashort Pulses and Applications
Coherent Phenomena
Coherent Sources for VUV- and Soft X-Ray-Radiation
Optical Storage and Memory
Tunable Lasers for Spectroscopy
Topics in Nonlinear Optics
Frequency Stable Lasers and Applications
Interaction of Laser Light with Matter
Mechanical Action of Laser Light
New Solid State Lasers
Laser Monitoring of the Atmosphere
Optical Bistability
Vapour Deposition of Material on Surfaces
Optical Computers
Phase Conjugation
Laser Spectroscopy
Multiphoton Ionization of Atoms and Molecules
Squeezed States of Light